The Permission to Be Right

How Knowledge Advances — and Why It Sometimes Doesn't

Phani Kurada, PhD

A portion of author proceeds from this book is donated to *Samskrita Bharati Kendram, Littleton, MA.*

First published 2026

ISBN: 979-8-9961846-0-6

Published by Kurada Press, Littleton, Massachusetts

In loving memory of my father, Sri Venkatramaiah Kurada,
and for my mother, Smt. Syamalamba Kurada

Acknowledgments

This book grew from a verse I did not expect to recognize, and from the habit, formed in a molecular biology laboratory, of refusing to accept a residual that does not fit.

My wife, Aparna Kurada, introduced me to Sanskrit, and therefore, without knowing it, to this book. She read every draft and improved all of them. Her judgment is sharper than mine on the things that matter.

My children, Abhinav and Sruthi, inspire me to keep growing.

The Samskrita Bharati Kendram in Littleton, Massachusetts gave me the language this book needed. I arrived not knowing what I was looking for. The classes taught me to parse what I was reading, and in doing so, revealed something I had not expected: that the precision encoded in Sanskrit grammar, where structure carries meaning without loss across centuries, is not unlike what a strand of DNA does. Both are solutions to the same problem: how to preserve exact information across centuries by encoding it in structure itself. That recognition, more than anything else, is what made this book possible.

Kripa Shankar Shukla and K. V. Sarma's critical edition of the *Āryabhaṭīya*, published by the Indian National Science Academy in 1976, is the primary source this book stands on. Without their scholarship, this argument could not have been made.

I began by not quite believing the claims. I followed the thread anyway. These are the people who made it possible to follow it honestly.

A Note on Method

This book does not attempt to be an exhaustive survey of the scholarly literature on Indian mathematics and astronomy. The claims made here rest on established scholarship: the primary source is Kripa Shankar Shukla and K. V. Sarma's critical edition of the *Āryabhaṭīya* (Indian National Science Academy, 1976), and the historical and mathematical claims draw on the work of serious scholars in the history of Indian science. Readers who wish to go deeper will find a rich and sometimes contested literature waiting for them.

The focus of this book is different. It follows a specific historical thread and asks what can be learned by tracing it carefully, not to catalog achievements, but to understand how they were produced, how they were carried, and where they reached and where they stopped. The framework presented here emerges from that process.

The goal is not to prove every step with footnotes but to make the reasoning visible, so that a reader can follow it, test it, and apply it elsewhere. Where the argument is interpretive, I have tried to say so. Where the evidence is incomplete, I have tried to say that too. The intellectual honesty the book demands of its subject, it tries to demand of itself.

One practical note: Chapter Six works through Āryabhaṭa's correction procedures in prose. For readers who want to follow the mathematics step by step, the procedure is reconstructed both visually and numerically, with worked examples, at phanikurada.com/aryabhata.

Contents

Part One

How a tradition approached its own limit

Chapter One

The Verse That Started Everything

The first time someone told me that ancient Indian mathematicians had anticipated calculus by two or three centuries, I didn't believe them.

I had grown up knowing some of these names — Āryabhaṭa, zero from India, Sanskrit as one of the oldest languages in the world — the way you know the names of famous ancestors: confidently, without detail, without mechanism. I had also heard claims like this: proud claims, made at family gatherings or in the margins of conversations about culture and history. Some of it was true. Some was stretched past what the evidence could bear. I had learned, over time, to hold such claims at a careful distance. I did not dismiss them. This subject attracts motivated reasoning in both directions: the impulse to inflate, and the impulse to deflate. I have tried to apply the same scrutiny to claims I wanted to be true and to those I expected to be false. The reader will judge whether I succeeded.

What I was looking for was not validation but mechanism. Not the surface — the structure.

That habit came from science. I spent years in a molecular biology laboratory learning to ask not just what a cell does but how it does it: what machinery underlies the visible behavior, what the system looks like when you go one level deeper. A beautiful result was not enough. I wanted to know what produced it. That laboratory habit (to treat the pattern as a question rather than an answer, to look for the mechanism, and to refuse to accept a residual that does not fit) became the primary analytical lens of this book.

I brought the same instinct to Sanskrit. I had grown up with the sound of it before I had any access to its sense: the cadences of ritual, verses at the edges of daily life, syllables that marked certain moments as different from ordinary speech. I knew its cadence before

I knew its grammar. When I finally started learning it properly, in my late fifties, in weekend classes run by a local branch of Samskrita Bharati, what struck me first was not the beauty, though it is real, but something I had not expected: the engineering underneath the beauty. Every inflection carries information. Every compound word is a small logical structure. In its learned form, the language is built to carry meaning without loss: across time, through memory, through the minds of people who might not understand every word but can reproduce every sound exactly. I wanted to understand what produces that quality, not to appreciate it from outside but to see the mechanism from within.

That search led me away from the literary texts and toward mathematics: specifically to the *Līlāvatī*, one of the most celebrated mathematical works in the Sanskrit tradition. The *Līlāvatī* is full of verses like the one I am about to show you: mathematics compressed into memorable form, each carrying a rule within it.

The verse reads, at first glance, like poetry. But something stopped me before I had finished the first line. I knew this formula. Not from Sanskrit scholarship: from secondary school mathematics, decades earlier. The sum of numbers from one to n, n times n-plus-one divided by two. It had sat in my memory all those years as an algebraic expression, clean and abstract, belonging to no particular place. And here it was again, encoded in a twelfth-century Sanskrit verse, dressed in the grammar of a language I was only beginning to learn. This was the verse that stopped me, because I already knew it. What I did not know was that someone had sung it.

> सैकपदघ्नपदार्धमथैकाद्यङ्कयुतिः किल सङ्कलिताख्या ।
>
> सा द्वियुतेन पदेन विनिघ्नी स्यात् त्रिहृता खलु सङ्कलितैक्यम् ॥
>
> *saikapadaghnapadārdhamathaikādyaṅkayutiḥ kila saṅkalitākhyā |*
>
> *sā dviyutena padena vinighnī syāt trihṛtā khalu saṅkalitaikyam ||*
>
> *"Half the count, multiplied by the count increased by one, is the total of the numbers beginning with one, which is what is called the sum (**saṅkalitam**). That, multiplied by the count increased by two and divided by three, is the sum of sums (**saṅkalitaikyam**)."*
>
> — Bhāskara II, *Līlāvatī* (12th century). Translated by the author.

It is not poetry in the ornamental sense. To see what it actually does, start somewhere small. Add the numbers from one to ten: 1, 2, 3, all the way to 10. You could work through it by hand and arrive at 55. Or notice that ten paired with one makes eleven, nine paired with two makes eleven: five such pairs, each summing to eleven, giving you 55 in a single move. Now ask: what if the sequence ran to a hundred? Or a thousand? The step-by-step method grows with the sequence. A rule does not. The labor no longer scales with the list.

That rule, the sum of the sequence from one to n, is one of the two things this verse encodes. The second is the sum of all those sums: a level of abstraction above the first. In modern notation, the two results look like this:

Sum of 1 to n (saṅkalitam): $n/2 \times (n + 1)$

Sum of those sums (saṅkalitaikyam): $\text{saṅkalitam} \times (n + 2)/3$

Both rules. One verse. Twelfth century. Encoded not in symbols but in words.

The significance is not the formulae themselves: versions of the first had been known in India for over a millennium before Bhāskara. The significance is what the verse does to the act of computation: it replaces a procedure that grows with the problem with a single structured operation that does not. It gives a machine for generating answers, for any value of n, forever.

What stopped me, though, was not the mathematics. It was how the mathematics was carried.

Sa-eka: with one added. *Ghna*: multiplied by. *Hṛta*: divided by. There are no symbols here in the modern sense: no n, no fraction bars, no equals sign. The operations are carried inside the words themselves. Each word is both language and instruction. The language is not describing the mathematics from outside: it is performing it. For the purpose of carrying a specific rule accurately across time, grammar becomes notation. The verse is simultaneously a poem you can memorize and an algorithm you can execute.

A result encoded this way (in a memorable verse, in a language whose grammatical structures enforce precision) can be carried in a person's memory across centuries. This is not a workaround for the absence of symbolic notation. It is a different solution to the same problem: how do you ensure that a precise mathematical rule survives and travels accurately? Make the language itself do the work of the notation.

But where had Bhāskara learned this? The formula did not originate with him. When I traced it backward, what I found deepened everything I had already understood.

In 499 CE, Āryabhaṭa, working in Kusumapura (modern Patna) recorded the same result in a single compressed sutra in the *Āryabhaṭīya*. His formulation is more general: the sum of any arithmetic progression, of which the natural number sequence is a special case:

> इष्टं व्येकं दलितं सपूर्वमुत्तरगुणं समुखं मध्यम् ।
>
> इष्टगुणितमिष्टधनं त्वथवाद्यन्तं पदार्धहतम् ॥
>
> *iṣṭaṃ vyekaṃ dalitaṃ sapūrvam uttaraguṇaṃ samukhaṃ madhyam |*
>
> *iṣṭaguṇitam iṣṭadhanaṃ tvathavādyantaṃ padārdhahatam ||*
>
> ***"The desired count, less one, halved, multiplied by the common difference, and added to the first term, is the mean. The mean, multiplied by the count, is the sum. Or else, the first and last terms added together, multiplied by half the count, also gives the sum."***
>
> — Āryabhaṭa, *Āryabhaṭīya, Gaṇitapāda* (499 CE). Translated by the author.

Read the two verses together and the difference is immediate. Āryabhaṭa's sutra is dense: every word load-bearing, nothing explained, nothing offered by way of welcome. It is mathematics as compressed transmission: the minimum words required to carry the rule without error. More than a century later, Bhāskara I expanded the same sutra in his commentary on the *Āryabhaṭīya* into step-by-step procedural prose: patient, sequential, executable. The mathematics is unchanged, but made visible to a reader who needed to follow it.

What emerges most clearly in Bhāskara II — what neither the sutra nor the commentary quite had — is a human presence on the other side of the mathematics. Someone being addressed. Someone for whom the problems are made alive.

Seven centuries. Three men. The same formula carried forward each time in a different form — compressed, then unpacked, then humanized: whether through direct transmission or through the tradition's deep commentarial culture, where each generation read and extended what the previous one had left.

The text is traditionally associated with Bhāskara's daughter. The problems are framed as if addressed to her, a young woman learning mathematics through problems about lotus pools and honeybees and merchants calculating their profits. Whether this framing is literally true, a later tradition, or a literary convention, the tone is unmistakable. This is mathematics taught with tenderness. A father teaching his daughter to see the structure underneath the world.

As a father, that stopped me completely.

I had come looking for mechanism and found, underneath the mechanism, a relationship. Precision and warmth, not opposed but reinforcing each other.

Something opened in me when I understood that. This tradition I had grown up adjacent to, had heard described in general terms of greatness without ever quite seeing what the greatness consisted of: it suddenly had a face. Not a monument. A father and a daughter, and the mathematics between them.

What followed was not a research plan, but a question: if a civilization encodes its mathematics in language precise enough to function as notation, what does that precision reveal about how the tradition actually worked: how it reasoned, how it composed and transmitted, and where it reached and where it stopped? And what happens to that knowledge when it travels, or when it does not? But the question underneath the question turned out to be larger than Indian astronomy: what does any tradition need to let its thinkers follow the evidence without fear, and what happens when those conditions are absent?

Āryabhaṭa, setting down his astronomy at twenty-three in a world of sacred cosmology, was another version of this question, a young man encoding precise observation in a tradition that appears to have let him replace a demon with a shadow without fear. The Kerala mathematicians, whose 15th-century work appears to anticipate elements of what we now call calculus, were another: a school so local that its knowledge barely traveled beyond its own region for four centuries. The verse about triangular numbers was not just a starting point. It was a template.

What I found, following that thread, was not the triumphant story I might have told. It was something more interesting: a tradition of genuine achievement shaped by patterns that do not appear to be accidents — a distributed authority that argued fiercely without silencing, a tolerance for questions that resisted closure, a language that encoded precision in memory — patterns with consequences for who published, and when, and without fear.

I began by not quite believing the claims. I followed the thread anyway.

The verse is still where it started. But it is no longer just a verse. It is the first piece of evidence in an argument I did not know I was making, about what a tradition looks like when you stop admiring it from a distance and start asking how it actually works. The tradition I examine here is Indian. But the question is larger than that.

Chapter Two

The Permission to Be Right

Āryabhaṭa, Galileo, Darwin, and What a Civilization Owes Its Thinkers

The verse had given me a method. What I needed now was a test.

Bhāskara II was the discovery, the moment the tradition revealed itself as something more than I had expected. But Bhāskara worked inside a world that had already been shaped by someone else. Seven centuries earlier, that world had faced a harder test: a young man who did not inherit permission but had to claim it. A man who walked into a cosmos run by demons and replaced them, calmly and under his own name, with geometry.

I had mentioned Āryabhaṭa at the end of the last chapter as a version of the same question the verse had planted. Now I want to examine him properly. Not just what he discovered, though that is remarkable enough, but what his story reveals about the structural conditions surrounding him, because those conditions are the argument.

To see those conditions clearly, I found I needed a mirror. Two mirrors: men from other traditions, equally brilliant and equally right, who fared very differently when they told the world what they knew.

> को अद्धा वेद क इह प्र वोचत् कुत आजाता कुत इयं विसृष्टिः।
>
> अर्वाग्देवा अस्य विसर्जनेनाथा को वेद यत आबभूव।।

> *ko addhā veda ka iha pra vocat kuta ājātā kuta iyaṃ visṛṣṭiḥ*
>
> |
>
> *arvāgdevā asya visarjanenāthā ko veda yata ābabhūva ||*
>
> *Who truly knows? Who here can declare it — whence it was born, whence this pouring-forth?*
>
> *The gods are later than this world's outpouring.*
>
> *Who, then, knows from what it came to be?*
>
> — Nasadiya Sukta, Rigveda 10.129.6. Translated by the author.

This is no peripheral verse. It stands within the Rigveda itself, among the foundational texts of the Vedic tradition. And it ends not with an answer but with a question. Not with certainty but with acknowledged uncertainty. Not with the gods confirming the official account but with the gods themselves placed after the fact: arriving too late to have witnessed the beginning, and therefore unable to tell us what happened.

That such a text sits inside the tradition's foundational inheritance is evidence of an epistemic disposition in which not knowing is not shameful, the honest admission of uncertainty is not weakness but wisdom, and the question is as sacred as the answer.

The Young Man Who Set Down Everything

To understand what Āryabhaṭa did, you must first understand what he was walking into.

Fifth-century India was not a secular environment. It was a world saturated with religious cosmology: ancient, authoritative, and woven into the fabric of daily life. The tradition was sophisticated: Indian astronomers had long understood the earth as a sphere: the very word bhugola, meaning earth-ball, was embedded in the astronomical vocabulary centuries before Āryabhaṭa. The Siddhāntic astronomical tradition already treated the earth as a globe suspended in space. On this, the tradition was ahead of much of the ancient world.

But that spherical earth sat motionless at the center of a revolving cosmos. The sun, moon, and stars rotated around it. And eclipses (those sudden, terrifying darkenings of the sky) were caused by Rahu, a celestial demon who periodically swallowed the sun and moon. This was not peripheral folk belief held loosely by the uneducated. It was present in texts of enormous authority, embedded in the cosmology that gave the world its meaning, and defended by scholars of considerable learning.

Into this world, in 499 CE, a twenty-three-year-old scholar in the city of Kusumapura (modern Patna) completed a work he called the *Āryabhaṭīya*. Shukla and Sarma, in their critical edition, suggest he was possibly the Kulapati (the head) of a scholarly institution there, or closely associated with one. Nālandā, nearby and emerging as a major scholarly center in this period, formed part of the broader institutional landscape of the region: a world where debate was the expected mode of intellectual life, and where authority was distributed across scholars rather than concentrated in a single doctrinal body. He wrote his treatise in verse, in Sanskrit, under his own name.

Among its claims:

> The earth turns: the apparent motion of the stars is not the stars moving, but the earth itself rotating beneath them. He expressed it with a vivid analogy: just as a man in a moving boat sees stationary objects on the shore moving backward, so the stationary stars appear to move westward for those on a rotating earth.
>
> Eclipses are not Rahu. They are shadows. The moon passes through the earth's shadow, or the moon's shadow falls on the earth. Geometry, not demons.
>
> The moon does not shine. It is a dark body illuminated by the sun, visible only because it reflects borrowed light. The lunar deity carries no light of its own.

> The value of pi is approximately 3.1416, one of the most precise approximations of the ancient world.

The spherical earth was accepted, but Āryabhaṭa was now asserting that this sphere spun. That the demon Rahu had nothing to do with eclipses. That the moon was a dark body. In crucial places, his claims did not merely supplement the tradition: they replaced its mechanics entirely.

He set them down. At twenty-three. Under his own name. Without apology.

And what happened next is the chapter's central question. Varāhamihira (d. 587 CE) attacked the rotation claim on physical grounds within decades of Āryabhaṭa's work. Brahmagupta, writing a century and a half after Āryabhaṭa, attacked it again — sharply, publicly, sometimes caustically — arguing that objects would fly off a spinning earth. His influence was so strong that even some followers of Āryabhaṭa eventually hedged on the rotation claim, and the debate split Indian astronomy into distinct schools: the *Ārya-pakṣa* following Āryabhaṭa, the *Brāhma-pakṣa* following Brahmagupta. Not every one of Āryabhaṭa's claims prevailed.

Notice what neither of them attacked. Neither Varāhamihira nor Brahmagupta objected to Āryabhaṭa's most theologically provocative claims. He said Rahu (a named demon in the Vedic cosmology) had nothing to do with eclipses. He said Chandra (the moon, worshipped as a deity in the Vedic tradition, present in ritual and prayer across the culture) was a dark body that borrowed its light from the sun. Not a god. A dark body lit by the Sun. Neither claim drew a religious objection in the surviving record. Neither Varāhamihira nor Brahmagupta mounted a theological defense of Rahu or Chandra. What they contested was the physics of a spinning earth, a question with no religious stakes at all. This distinction is the chapter's central point: on questions of physical cosmology, the tradition's debates were scientific, not theological. The silence around Rahu and Chandra is as revealing as the debate about rotation. If the tradition had a mechanism for punishing heterodox claims, those are where it would have activated. They did not.

To be argued against is not the same as to be silenced. One preserves the possibility of being right; the other forecloses it. Āryabhaṭa's rotation claim was set aside by reasoning that lacked the physics to complete itself, not by a verdict that placed it beyond reach. That no verdict could be issued was the tradition's great protection.

What demands explanation is not the astronomy. It is that the astronomy could be stated at all: without apology, under his own name, without fear of institutional punishment. The explanation is not personal courage alone. It is structural: no single institution had both the power and the incentive to silence him.

The Man Who Had to Unsay It

In 1633, Galileo Galilei stood before the Inquisition in Rome and recanted. He stated, formally and on record, that the earth did not move around the sun: a statement he knew to be false, that the evidence had made increasingly difficult to defend on astronomical grounds, and that everyone in the room understood was not really about astronomy at all.

It was about authority.

The Church had not arrived at geocentrism through careful astronomical observation. It had inherited a cosmological framework from Aristotle and Ptolemy, found that it did not obviously contradict scripture, and over centuries allowed that framework to become entangled with doctrine, with institutional prestige, with the social architecture of a civilization that had placed the Church at the center of all knowledge. When Galileo turned his telescope on Venus and found phases like the moon's, and on Jupiter and found four moons orbiting it, he had evidence the Ptolemaic system could not absorb. By the time he stood before the Inquisition, the earth's position in the cosmos was not a scientific question. It was a political one.

To say the earth moves was to say the Church was wrong. And to say the Church was wrong was to say the institution that claimed authority over all truth — spiritual, moral, natural — had been teaching error for centuries. It threatened not cosmology but institutional legitimacy.

This is the machinery that produced the recantation. Not malice exactly, though there was malice. Not stupidity, though there was willful blindness. But something more structural and therefore more dangerous: an institution that had staked its authority on a specific answer, and had built no mechanism for being wrong gracefully. The same institutional landscape could work in the opposite direction, but not neutrally. Kepler: the German mathematician and astronomer whose planetary work is examined in detail later in this book: pursued his planetary work under the protection of Rudolf II, Holy Roman Emperor, who collected exceptional minds as expressions of imperial prestige. The patronage was real and sustained. But it was instrumental: Kepler was protected not because heterodox science was valued for its own sake, but because Rudolf's ambitions made it useful. Authority shapes knowledge in both directions: as censor and as engine, and in both cases, the direction depends not on the truth of the claim but on whose agenda it serves.

Galileo spent the remaining nine years of his life under house arrest. He continued working: his final book, written in secret and smuggled out of Italy, contained some of his most important physics. But he could not publish under his own name in his own country. He could not stand in the sunlight of his own ideas.

He had been given the opposite of permission.

The Man Who Waited Twenty Years

In 1838, two years after returning from his voyage on the Beagle, Charles Darwin worked out in his private notebook the theory that would eventually become On the Origin of Species.

He spent the next twenty years, twenty years, accumulating evidence and waiting.

He knew that to declare his conclusions was to declare war on beliefs held not as opinion but as identity. He wrote to his friend Joseph Hooker in 1844: six years after completing the theory, that admitting them felt like "confessing a murder."

His case differs in form from Galileo's — there was no tribunal, no formal recantation demanded — but not in structure. His claim confronted a social and religious order deeply invested in a particular account of human origins, and the consequence was years of delay rather than open transmission.

He finally published in 1859, forced into it by Alfred Russel Wallace independently arriving at the same theory. Twenty-one years after he had worked it out.

Āryabhaṭa did not wait. He composed and transmitted at twenty-three.

Conditions, Not Destiny

What determines whether a truth can be stated without fear is not its subject matter. It is the structure it encounters. Where an authority exists that has both the power to define orthodoxy and the incentive to defend it, disagreement can be converted into punishment: regardless of whether the claim concerns the motion of the earth, the origin of species, or anything else that touches what the authority has staked its legitimacy on. Where no such authority exists, disagreement remains what it should be: argument. Permission becomes visible at the point of disagreement, when the result challenges something an institution has reason to defend. The cases this chapter has examined differ in domain and in century. What they share is a structural condition, or its absence.

The Vedic intellectual tradition had no single central authority with both the power and the mandate to enforce one answer to questions about the physical world. Competing schools, commentators, and traditions disagreed constantly and openly. The Buddhists and the Jainas argued entirely outside the Vedic framework. Brahmagupta could attack Āryabhaṭa as vigorously as he liked, and he did, but he had only the force of his arguments. No council could convene and issue a verdict.

Darwin did not lack courage. Galileo did not lack brilliance. What they lacked was the structural assurance that following the evidence was safe, the assurance that comes only when no single institution appears to have held both the power and the incentive to close a question permanently. Āryabhaṭa had that assurance not because his tradition lacked conflict, but because its conflicts could not become verdicts.

That is what cultural permission looks like when it works.

Permission is not the same as agreement. Brahmagupta disagreed with Āryabhaṭa vigorously: fiercely even, and that disagreement produced distinct schools of astronomy that sharpened both traditions. The tradition had disagreement, but not disagreement that prevented coexistence. Brahmagupta and Āryabhaṭa could both be right and both be wrong simultaneously, their schools operating in parallel for centuries, each refining its own parameters without being forced to destroy the other. That is absorptive pressure. It sharpens. It does not break. Criticism is not the enemy of inquiry. It is one of its primary engines. What a civilization owes its thinkers is not protection from argument but protection from punishment.

Darwin released his work twenty-one years late. Galileo wrote his claims down and was forced to publicly disown them. Āryabhaṭa composed and transmitted at twenty-three, under his own name, having just replaced a demon with a shadow and a dark body with a borrowed light, and the worst that happened was that someone wrote a fierce rebuttal.

The genius was equally distributed. The permission was not.

Each of these men held a finding that contradicted what the prevailing framework said should be true. An anomaly: a result that does not fit, that the system has to do something with. Āryabhaṭa's eclipse claim was correct. Darwin's theory was correct. Galileo's evidence was correct. What happened to each anomaly — absorbed through argument, delayed by social danger, suppressed by institutional verdict — was determined not by what was found but by what the surrounding structure was built to do with it.

Whether a structure absorbs an anomaly or is forced to resolve it — what makes the difference between a contested position that persists for centuries and a finding that forces a framework to break — is the prior question. It is also the question that the most serious accounts of how knowledge systems change have left open.

The same structural condition that gave Āryabhaṭa the freedom to state what he knew is also the condition that allowed the tradition to absorb what it found, without being forced to complete what it had started. The permission to be uncertain cuts both ways. It protects the thinker from punishment. It also means the tradition never has to resolve what it can afford to leave open. The freedom and the limit are not opposites. They are the same structure, seen from different angles at different moments.

This argument sits in a conversation that deserves to be named. Thomas Kuhn's account of scientific revolutions describes what happens when anomalies accumulate into crisis: normal science is disrupted, a new paradigm emerges, and the community eventually converts. The description is powerful. But Kuhn does not explain what makes the crisis unavoidable: what determines when anomalies can no longer be ignored.

Imre Lakatos moved closer. His account of research programs describes how systems protect their core assumptions: a surrounding layer of adjustable assumptions absorbs anomalies, adjusting parameters rather than surrendering the framework. Lakatos explains resistance. What he does not specify is when that resistance fails.

Nassim Taleb's work on fragility describes what happens when systems break under stress. But it does not distinguish between systems that can absorb stress indefinitely and those that cannot.

What these accounts leave open is the condition under which change becomes unavoidable.

Anomalies do not force change simply by existing. Change occurs only when an anomaly becomes inescapable. For that to happen, three conditions must align. The anomaly must be visible. It must persist in a way that resists explanation. And the system must be structured so that it cannot indefinitely defer it. When these conditions come together, the anomaly becomes a verdict. When they do not, the system continues.

The two traditions this book examines demonstrate the distinction. Kerala provides a case of the former: anomalies visible, mathematicians capable, insight present, and the system continuing. Kepler's Europe provides a case of the latter: the same class of anomaly, under different structural conditions, producing rupture.

The difference is not the anomaly itself. It is the system's capacity to absorb it.

Āryabhaṭa's case demonstrates the framework in miniature. He had an anomaly: a cosmos explicable by geometry rather than by demons. He could state it because no single institution held both the power and the mandate to close the question. The system absorbed the claim. It argued back. It did not rupture.

को अद्धा वेद — Who truly knows?

That question, placed at the heart of one of the world's oldest continuously transmitted literary traditions, is not an expression of despair. It is an expression of intellectual honesty so deep it became sacred. A tradition that could ask that question about the origins of the cosmos, that could look at its own foundational mystery and say, with full seriousness, we do not know, and neither do the gods — had already created conditions in which authority could not manufacture certainty where none existed.

The permission to be right begins with the permission to be uncertain. A culture that cannot tolerate not knowing tends, over time, to find ways of suppressing those who know, not always through explicit decree, but through the slower mechanisms of professional exclusion, social pressure, and the hedging that makes honest people sit on honest ideas for twenty years.

There was no single throne from which to resist.

♦

I have been sitting with Āryabhaṭa's story for a long time now, and what stays with me is not the astronomical precision, remarkable as it is. It is the image of a young man (twenty-three years old, the same age I was at the start of my own scientific training in India) who simply wrote down what he had worked out and gave it to the world.

What his world had given him was something more valuable than encouragement: the structural conditions in which his honesty could not be used against him.

The verse about triangular numbers — the sums of sequences from one to n, each forming a triangle in dot form, encoded in a twelfth-century Sanskrit verse — led me here. And here, I began to understand that the tradition I had been studying was not just a repository of achievements but a structure shaped by recognizable conditions — distributed authority, tolerance for unresolved questions, argument without verdict — conditions whose presence in the institutional world Āryabhaṭa inhabited, and whose consequences in his career, are visible in the record, even if the record is incomplete.

But understanding what a tradition achieved, and under what conditions, is only the beginning of honest inquiry. The harder question, and the more necessary one, is how we evaluate such claims with the rigor they deserve. What counts as evidence? What does it mean when two traditions arrive at similar results independently? And what happens when the record is genuinely incomplete, when we can see the sender and the destination but not the path between them?

Those are not rhetorical questions. They are the method. And the method matters here because the claims already made in this book are themselves historical claims: about what the tradition achieved, about the conditions that surrounded it. Evaluating them honestly requires a standard of evidence. What that standard looks like, and how to hold it consistently in all directions, is what comes next.

Chapter Three

Equivalence Is Not Evidence

Epicycles, Calculus, and the Illusion of Borrowing

A recurring argument in the history of astronomy holds that Āryabhaṭa's planetary correction system — his *manda* (slow) and *shīghra* (fast) adjustments for planetary irregularities, examined in detail in later chapters — was derived from Ptolemaic epicycles. The reasoning runs as follows: the two systems produce mathematically equivalent results; Greek astronomy predates Āryabhaṭa; therefore, Indian astronomy must have borrowed from Greek models.

This argument has serious scholars behind it. B. L. van der Waerden, a historian of mathematics whose work commands genuine respect, argued that the structural similarities between Indian and Greek planetary models reflect direct transmission, that Āryabhaṭa's corrections were ultimately derived from pre-Ptolemaic Greek epicyclic thinking. The argument is not casual. It is careful, detailed, and made in good faith.

It is also, on examination, insufficient. And the reason it is insufficient reveals something important about how historical arguments of this kind work, and fail.

Noel Swerdlow, examining the same evidence van der Waerden examined, found his inferences to be in direct contradiction of the texts. Two careful scholars, looking at the same mathematical relationships, reached opposite conclusions. That outcome is not a failure of scholarship. It is exactly what we should expect when mathematical equivalence is made to carry the weight of historical evidence. Equivalence can support a transmission claim. It cannot establish one. The gap between those two things is the subject of this chapter.

More importantly, equivalence has been treated inconsistently, accepted as sufficient evidence in some cases and rejected as insufficient in others. That inconsistency is the methodological problem: it is directional, not neutral, and it consistently operates in one direction.

If the asymmetry reflected evidence availability, we would expect the standard to tighten or loosen case by case. What we observe instead is a consistent directional pattern.

What Equivalence Tells Us

Transmission is a matter of historical record. It requires named intermediaries, datable pathways, and evidence that one tradition had access to another.

Equivalence is a property of mathematics. When two systems produce similar results, that tells us something real: that both traditions were solving the same problem under similar constraints, and that this structure pushes careful minds toward similar solutions. It does not tell us who learned what from whom.

Consider the specifics of the Ptolemy-Āryabhaṭa case. Ptolemy's system models planetary motion using explicit geometric constructions: deferents and epicycles, circles upon circles, with fixed radii arranged to reproduce the observed irregularities in planetary paths. An epicycle, to give the reader a foothold, is a small circle whose center travels along a larger circle called the deferent, itself offset from Earth rather than centered on it, together tracing the irregular path a planet appears to follow across the sky. Āryabhaṭa's system operates through anomaly-based corrections expressed as computational procedures applied to angular quantities. At first glance the systems look analogous. On closer inspection the differences are structural and fundamental.

Ptolemy's epicycles are geometrically rigid: fixed circles of constant radius tracing fixed paths. The Indian system, as reconstructed in modern analysis, involves effectively varying parameters: producing motion that, for small variations, approximates elliptical behavior, though not described that way in the original framework. This feature has no clear counterpart anywhere in Greco-Roman astronomical texts. Ptolemy himself, struggling to reconcile his geometric constructions with observational data, introduced the equant as a workaround, an acknowledgment that his own system's physical coherence was already strained. The Indian system addressed related problems through a structurally different mechanism. And there is a further point worth noting: Āryabhaṭa deployed his planetary corrections within a cosmological framework that included a rotating Earth, a claim the Greek epicyclists did not make and would not have accepted. When a tool is adapted to serve a purpose its supposed originators explicitly rejected, the argument for borrowing weakens considerably.

Modern analysis can map these two approaches onto each other, but only after translating both into a shared mathematical framework that neither tradition originally used. What appears to us as "the same" is the result of retrospective reinterpretation. The similarity is real. But it is a similarity we are projecting backward, not one the original practitioners would have recognized.

The Third Possibility

Ruling out borrowing does not rule out connection.

The ancient world was not hermetically sealed. Trade routes connected the Roman and Byzantine world to Gupta India. Ideas and problems could travel without intact solutions traveling with them.

Historians call this stimulus diffusion: shared intellectual pressure generating structurally distinct local solutions. But it too requires evidence: records of contact, shared problem contexts, a demonstrable chronological gap. Without these, it remains a possibility, not a finding.

The Same Logic, Run in Reverse

There is a figure we have already met in this book who sharpens the argument considerably.

In Chapter One, I encountered Bhāskara II through a verse in the *Līlāvatī*: a formula for the sum of sequences that I recognized from secondary school mathematics, now revealed to have been sung six centuries before I learned it. What that chapter did not mention is that Bhāskara's mathematical work extends well beyond the arithmetic of sequences. He developed, in the 12th century, ideas that resemble what we would later call the derivative of the sine function, and articulated concepts about instantaneous velocity that bear recognizable resemblance to elements later formalized in the differential calculus of the 17th century.

The mathematical resemblance is real. The chronological priority of the Indian work is real. Bhāskara predates Newton by five centuries and Leibniz by roughly the same. By the reasoning applied in the Āryabhaṭa-Ptolemy case — equivalence plus chronological priority equals probable transmission — one could construct an argument that Newton borrowed from Bhāskara II.

No serious historian makes that argument.

Not because the logic fails in one case and holds in the other, but because the standard is applied differently when the direction of transmission reverses.

And rightly so. There is no documented pathway from 12th-century Indian mathematics to 17th-century European calculus. No known Latin translations of Bhāskara's work circulated in Newton's world. No intermediary has been identified who could have carried the specific mathematical ideas across the gap. In the absence of that record, independent development is not a concession: it is the honest position.

But notice what has happened. The same logical structure — mathematical equivalence plus chronological priority — that is treated as sufficient evidence when India is the borrower becomes insufficient when Europe might be. The standard shifts, and not randomly.

The Asymmetry

The asymmetry is methodological, not incidental.

Growing up in post-independence India, I received an education that gave me Marconi but not Jagadish Chandra Bose, and this was not British policy. The British had long since left. It was the curriculum of an independent nation that had not yet decided what to reclaim from what had been set aside. Bose had publicly demonstrated wireless signaling in Calcutta in 1895, before Marconi's public London demonstrations in 1896; a mercury-coherer detector derived from Bose's work was used by Marconi in 1901 to receive the first transatlantic wireless signal; and IEEE has recognized Bose as one of the fathers of radio science, commemorating his 1894–96 millimeter-wave communication experiments as an IEEE Milestone. He appeared in my schooling, if at all, as a footnote. The default assumption, absorbed rather than stated, was that the direction of knowledge ran from the West inward, and that Indian contributions were local variations on a universal story that originated elsewhere.

The asymmetry in historical scholarship that I am describing is the academic version of the same assumption. When Indian results resemble Greek ones, transmission from Greece is readily inferred: the priority of the older tradition treated as sufficient grounds for assuming the direction of influence. When Indian results resemble or predate European ones, independent discovery is assumed without argument. The possibility of transmission in the other direction is not even seriously entertained.

This inconsistency is not a minor scholarly quibble. It reflects a deeper methodological failure: the selective application of evidentiary standards depending on the direction of the proposed transmission. A methodology that demands proof in one direction while assuming it in the other is not a methodology. It is a prior conclusion dressed in the language of scholarship.

The correction is not to reverse the asymmetry. It is to remove it. Apply the same standard in all directions.

What the Standard Actually Requires

Mathematical equivalence, discovered after the fact, cannot serve as evidence of historical influence. Āryabhaṭa may have encountered Greek ideas through the channels of exchange that undoubtedly existed between the Hellenistic world and Gupta India. Features of the Indian planetary model appear to be pre-Ptolemaic, leaving the direction of any influence, if it existed at all, genuinely unresolved.

What we do know is that the channel existed and that ideas traveled through it. The most consequential example is the numeral system itself. The place-value numerals that Āryabhaṭa used for his astronomical calculations (the system that makes large number computation tractable, that underlies every calculation in the *Āryabhaṭīya* and every computation the Kerala School subsequently performed) are Indian in origin. They traveled westward through Arab intermediaries, reaching Europe by the 9th and 10th centuries, where they displaced the Roman numeral system and made modern mathematics possible. That transmission is documented: we have the manuscripts, the named intermediaries, the datable texts. The channel between the Indian mathematical tradition and the West was not hypothetical. It carried the most important mathematical tool in history. The specific question of whether it also carried the Kerala calculus remains unresolved. But the existence of a functioning channel is not in dispute.

To establish it, one must find traces in the historical record: named intermediaries, datable manuscripts, documented pathways of exchange, recognizable transformations of one system into another as it crossed a cultural boundary. When those traces exist, transmission can be established. When they do not, the honest position is to note their absence and resist the temptation to fill the gap with assumption.

Transmission requires a chain of custody: identifiable intermediaries, datable texts, and traceable transformations linking source to destination. That chain either exists in the record or it does not.

This is a demanding standard. It is also a fair one. It is the same standard we would apply to any claim about knowledge moving between any two cultures, in any direction, at any period of history. Holding it consistently is not skepticism about Indian achievement: it is respect for it. A tradition that genuinely produced remarkable results does not need inflated claims. Inflated claims, when they fail under scrutiny, damage the credibility of the genuine achievements they were meant to protect.

I found this standard clarifying rather than deflating. What I wanted, following the thread, was not to arrive at a conclusion I had decided in advance. I wanted to know what the evidence actually showed, even when, as here, it showed less than the proud claim required. That restraint is not skepticism about the tradition. It is the only posture that leaves the genuine achievements standing on ground that cannot be taken away.

Equivalence reveals the unity of mathematics. It does not reveal the path of ideas. That path, when it can be established at all, requires different evidence entirely.

Confusing the two produces conclusions that feel plausible but cannot be defended.

♦

The verse in Chapter One stopped me because I recognized the formula. But tracing it backward (through Bhāskara I's commentary, back to Āryabhaṭa's compressed sutra) was an act of following a thread without knowing where it led. At no point did I need to claim that the tradition borrowed nothing and gave everything. I needed only to ask what the evidence actually showed, and to hold that question open when the evidence was incomplete.

The next chapter examines what it looks like when the standard is actually met.

Chapter Four

When Equivalence Does Suggest Transmission

The Case of Hindu-Arabic Numerals

The previous chapter established a standard. This one shows the standard being met in full.

Everything argued there — that equivalence is insufficient, that transmission requires documentary evidence, that the standard must hold in all directions — was a negative claim. It said what evidence cannot do. The contrast with the previous chapter is deliberate: where equivalence fails, documented transmission leaves a trace. And when that trace exists, the question is not open.

The answer, in the history of mathematics, has a name most people encounter every day without knowing its origin. It is written in the numerals you used this morning to check the time, read a price, or dial a number. And the story of how those numerals traveled from India to the Islamic world and then to Europe is one of the cleanest cases of documented transmission in the history of mathematics. Examining it carefully is not merely an exercise in appreciation. It is a demonstration of what the standard requires, and what it looks like when the evidence actually meets it.

The Numerals We Use

The symbols 0 through 9, arranged in the positional decimal system that makes modern arithmetic possible, originated in India. This is not a disputed claim. It is documented history. The system developed over several centuries from Brahmi numerals into the positional notation we recognize today, a notation whose key innovation is not the symbols themselves but the principle that a digit's value depends on its position. The 3 in 300 is worth one hundred times the 3 in 3. That positional principle, which we now take so entirely for granted that it is invisible, was a genuine intellectual achievement. It made calculation tractable in ways that additive systems like Roman numerals simply cannot match.

The transmission of this system westward was not inferred from mathematical similarity. It was documented at every stage by named individuals who explicitly credited their sources. This is the first and most important thing to understand about the Hindu-Arabic numeral case: the evidence for transmission is not circumstantial. It is direct. It consists of people saying, in writing, where they learned what they knew.

The Chain of Custody

The first critical link is Al-Khwārizmī, a Persian mathematician working at the House of Wisdom in Baghdad around 825 CE. The House of Wisdom was the great intellectual institution of the Abbasid Caliphate: a center of translation, scholarship, and original research that drew thinkers from across the known world. Al-Khwārizmī was among its most consequential figures. He wrote a treatise whose title announces its origin without ambiguity: On the Calculation with Hindu Numerals. The Arabic original is lost, but a 12th-century Latin translation survives. The work's influence was so wide and so lasting that Al-Khwārizmī's name, Latinized as Algoritmi, became the root of the word algorithm.

He does not merely use the numerals. He names their origin explicitly. This is what it looks like when transmission is not inferred but demonstrated.

The second link is Al-Bīrūnī, the 11th-century polymath who spent more than a decade in India and did something that still commands respect a thousand years later: he learned Sanskrit. Not casually, not at a distance, but well enough to read original Indian scientific and mathematical texts and translate them systematically for Arabic-speaking readers. His Kitāb fī Tahṣīl mā lil-Hind (known in English as the Indica) is a remarkable work of comparative scholarship, engaging Indian mathematics, astronomy, philosophy, and cosmology with genuine depth. What it represents is not admiration from a distance but direct linguistic access: the kind of engagement that leaves a documentary trail.

This is not transmission inferred from similarity. It is transmission recorded.

The third link is Fibonacci (Leonardo of Pisa) who in 1202 published the Liber Abaci, which helped bring the Hindu-Arabic numeral system into wider use in Latin Europe. He called the numerals figuras Indorum, the figures of the Indians. He credited their origin explicitly. There is no ambiguity. The source is named. He had encountered them in North Africa, through Arabic intermediaries, while traveling as a young man with his merchant father. The chain, India to the Islamic world to Latin Europe, runs through documented individuals, named texts, and explicit acknowledgments of debt across four centuries.

The Book That Traveled

There is a connection to an earlier figure in this book that deserves to be named directly.

In Chapter Two, Brahmagupta appeared as Āryabhaṭa's fiercest critic: the man who argued publicly and caustically that objects would fly off a spinning earth, whose influence was strong enough to split Indian astronomy into rival schools for centuries. In Chapter Three, his name appeared again: certain features of the Indian planetary model that Brahmagupta worked within were noted as potentially pre-Ptolemaic, complicating the borrowing argument in ways that remain unresolved.

What neither of those chapters mentioned is what happened to Brahmagupta's own work after his death. His great mathematical and astronomical text, the *Brāhmasphuṭasiddhānta*, completed in 628 CE, was the book that scholars from Sindh carried as emissaries to

Baghdad in the 8th century. It was received at the court of the Abbasid Caliph Al-Mansūr and translated into Arabic under the title *Sindhind*. This was not incidental. This is transmission with a traceable event: a text carried, translated, and absorbed into another tradition. It was one of the earliest and most significant acts of scientific translation in the history of Islamic scholarship, the moment when Indian mathematical astronomy entered the Arabic intellectual world and began the journey that would eventually reach Fibonacci and Latin Europe.

The *Sindhind* translation is transmission embedded in political and diplomatic history: scholars arrived at a court, they carried books, the books were translated. We know when, and roughly why. That is the kind of record that makes a historical claim durable.

The Standard, Demonstrated

Chapter Three argued that equivalence cannot establish transmission. This chapter shows what can. The Hindu-Arabic numeral case leaves no room for inference. The chain is visible at every step. This is not inference. It is record. The central question is not in serious doubt. The links are: named intermediaries, explicit acknowledgment of sources, datable texts, documented institutional pathways, and traceable structural transformations.

The two chapters together make a single methodological argument. Hold the evidentiary standard consistently, in all directions, for all traditions. Where the evidence meets the standard, affirm transmission. Where it does not, decline to infer it. This is not skepticism. It is the only approach that holds, the only one that protects both the genuine achievements of the tradition and the credibility of the claims made on their behalf.

The history of mathematics is rich enough to survive honest accounting. The Hindu-Arabic numerals traveled from India to Baghdad to Fibonacci's *Liber Abaci* to the world, and we can trace every step of the journey because the people who carried them said so. That is a remarkable story. It does not need inflation. It needs only to be told accurately.

And because the evidence is complete, the conclusion does not depend on interpretation.

♦

There is something quietly extraordinary about the word *algorithm.* It sits inside the vocabulary of the most technologically advanced civilization in human history, embedded in the logic of search engines and machine learning and satellite navigation, and it carries inside it the name of a man in ninth-century Baghdad who was careful enough to say where he had learned what he knew. Al-Khwārizmī did not need to name his source. He chose to. And that choice, made in a specific city in a specific year for reasons we can only partly reconstruct, left a trace that survived a thousand years.

The Hindu-Arabic numeral case is the cleanest case in this book. The evidence is complete, the chain is unbroken, and the standard is met at every link. What comes next is harder: a case where the sender is known, the channel existed, the timing is suggestive, but the receipt has never been found. A case where serious scholars have looked carefully at the same evidence and reached different conclusions.

That uncertainty is not a failure of the method. It is the method working.

Chapter Five

The Sender and the Channel

The Kerala School and the Hardest Question

The previous chapter showed the method working cleanly. A complete chain, unbroken from source to destination, documented at every link by people who said where they had learned what they knew. Al-Khwārizmī's title. Fibonacci's *figuras Indorum*. Scholars carrying the *Brāhmasphuṭasiddhānta* to Baghdad. The Hindu-Arabic numeral case is the kind of case that makes historical argument feel like detection: evidence gathers, the chain holds, and at the end you can say with confidence that this traveled, and here is how.

This chapter is different. It examines a case where the sender is known, where the channel plausibly existed, where the timing is suggestive enough to raise serious questions, and where the receipt has never been found. A case that is not pure equivalence, because the circumstantial evidence is real and historically grounded. But not documented transmission either, because the documentary record that would confirm it is absent.

This is the territory where most interesting historical questions actually live. Not the clean cases at either end: obvious equivalence on one side, obvious documentation on the other, but the difficult middle, where serious scholars have looked carefully at the same evidence and reached different conclusions.

The case is the Kerala School of mathematics. The question is whether its discoveries in infinite series, made in the 14th and 15th centuries, preceded and possibly influenced the development of calculus in Europe. And the honest answer, after following the evidence as carefully as I know how, is: we do not know. What we can do is map precisely what the evidence shows and what it does not, and understand why keeping the question open is not a failure of nerve but the only position the evidence supports.

The Achievement

In Chapter Three, Bhāskara II appeared briefly as a figure whose 12th-century work developed ideas resembling what we would later call the derivative of the sine function, a connection to calculus that predates Newton by five centuries. That chapter used Bhāskara to illustrate the asymmetry of evidentiary standards: the same logic applied in reverse would suggest Newton borrowed from him, a claim no serious historian makes.

Two centuries after Bhāskara II, in a small region on the southwestern coast of India that is now the state of Kerala, a mathematician named Mādhava of Saṅgamagrāma took the tradition considerably further. He did not merely approach calculus. He worked out, in the 14th century, results that are recognizable as foundational to it, not as approximations or precursors, but as genuine mathematical achievements that European mathematicians would not reach for another two to three centuries.

Mādhava derived infinite series expansions for trigonometric functions: the kind of series that allow you to compute the sine or cosine of any angle to any desired precision through addition alone, without tables. He developed methods the Kerala tradition would later use to compute π to seventeen decimal places. His successors, working through the 15th and 16th centuries in what became known as the Kerala School, developed systematic methods amounting to early forms of power series, iterative approximation, and the relationship between integration and summation. These results were recorded in Sanskrit texts: among them the *Tantrasaṅgraha* and the *Yuktibhāṣā*, the latter remarkable for providing detailed derivations and proofs rather than merely stating results.

The chronological priority of these results relative to Newton and Leibniz is not in dispute. Mādhava's infinite series for the sine and cosine anticipate by roughly three centuries the series that James Gregory and Leibniz developed independently in Europe. The question is not whether the Kerala School achieved these results. It did. The question is whether that achievement remained within Kerala, or whether it traveled.

The Evidence for Transmission

The case for transmission rests on several converging lines of circumstantial evidence. None is individually conclusive. Together, they create a pattern that cannot simply be dismissed.

The first is opportunity. Portuguese traders arrived in Kerala in 1498: the same Kerala where the mathematical school was active, in the same decades when its most sophisticated results were being worked out and recorded. Jesuit missionaries followed, establishing colleges, printing presses, and sustained institutional contact with local scholars. The Jesuit presence in Cochin, then the center of the Kerala mathematical tradition, was not incidental. It was organized, sustained, and included people with genuine intellectual interests.

Jesuit correspondence from the late 16th century records attempts to understand local timekeeping practices from Brahmin scholars in the Cochin region. The Kerala mathematical tradition was centered on precisely the astronomical and calendrical knowledge they were seeking. The correspondence does not mention infinite series or trigonometric calculations. But it places Jesuit scholars in intellectual contact with exactly the community that possessed them, at exactly the moment when that community's knowledge was most developed.

The second is motivation. Pope Gregory XIII's reform of the Julian calendar in the 1570s required improved astronomical knowledge. The Jesuit mathematician Christopher Clavius, a central figure in the reform, explicitly sought information on timekeeping practices across cultures. The practical problem — how to construct a more accurate calendar — was precisely the problem that Kerala's mathematical tradition was unusually well equipped to address. There was an institutional need on the European side and an intellectual tradition on the Indian side that could have met it. Whether those two things ever connected is a different question. But the motivation for connection was real.

The third is chronological suggestiveness. Infinite series expansions for trigonometric functions begin appearing in European mathematics in the early 17th century — in the work of Cavalieri, Fermat, Pascal, and Gregory — roughly a generation after sustained Jesuit presence in Kerala. This is not proof. A generation is a plausible transmission time; it is also a plausible period for independent development. The timing is consistent with transmission. It is equally consistent with coincidence. What it does is raise a question that cannot be answered by the timing alone.

The Evidence Against: A Comparative Silence

The case against transmission is equally serious. It rests not on counter-evidence but on a specific and significant absence: the absence of exactly the kind of record that, in the Hindu-Arabic numeral case, makes transmission undeniable.

Historians such as Kim Plofker and Ugo Baldini have examined the relevant archives extensively. Unlike the numeral case, where we have Al-Khwārizmī naming his source in his title, Al-Bīrūnī reading Sanskrit originals, Fibonacci calling them the figures of the Indians: there are no known Latin translations, summaries, or even references to Kerala mathematical texts in European sources of the 16th or 17th centuries. Not a single document records a European mathematician learning these series from Indian sources.

David Bressoud, one of the most careful scholars of this question, states the position plainly: there is no clear evidence the Kerala work on infinite series was known beyond India, or even beyond Kerala itself, until the 19th century. The narrowing geography of that claim is striking. The tradition appears to have remained not merely within India but within a single regional school: a community of scholars working in a small area of the southwestern coast, whose results did not travel even within India, let alone across the ocean to Europe.

This is the contrast with the numeral case in its starkest form. In the numeral case, we have the full chain: sender, channel, and receipt. In the Kerala case, we have the sender and the channel: Mādhava and his school, and the Jesuit presence in Cochin, but no receipt. What we lack is the confirming trace: the document that shows the knowledge crossing the gap.

There is a further consideration. The Jesuits present in Kerala were often in transit to other destinations, particularly China, and there is little evidence that they acquired the level of linguistic and mathematical expertise that engaging with a text like the *Yuktibhāṣā* would have required. Remember what it took for Al-Bīrūnī to engage with Indian mathematical texts: more than a decade in India, Sanskrit learned to the level of technical fluency, sustained scholarly immersion. The Jesuits in Cochin were missionaries and administrators, some with mathematical literacy, none with documented access to the Kerala tradition's most sophisticated results. The channel existed in the sense that contact existed. Whether it was the kind of channel through which advanced mathematics could travel is a different question.

There is a deeper methodological issue here that the transmission debate rarely addresses. Western historical method is calibrated for written records, physical manuscripts that can be dated, traced, compared, and verified by material analysis. The Indian mathematical tradition was not stored that way. It was stored in structured verse, encoded through systems like *kaṭapayādi* (a system Chapter Eight examines in full), where each consonant carries a numerical value and a verse encodes a precise mathematical constant in a form that self-corrects against degradation: alter one syllable and the encoded number breaks, making corruption immediately detectable. This is not folk memory. It is an encoded archival technology that happens to use human minds and grammatical structure as its medium rather than parchment and ink. When Jesuit missionaries encountered Kerala's astronomical tradition, they would not have found a library of manuscripts to copy. They would have found practitioners chanting mathematical verse.

The absence of a Latin translation in European archives remains decisive evidence under the standard this book applies. What this observation suggests is not that transmission occurred, but that the absence of documentary evidence may reflect both the limits of the surviving record and the limits of the method. The channel existed in a form the archive cannot verify. That is not a reason to lower the standard. It is a reason to name what the standard can and cannot see.

There is one further consideration that the transmission debate has not adequately weighed. In 1498, the same Portuguese expedition that arrived at Calicut carried an unnamed Gujarati pilot who navigated the Indian Ocean crossing that made the voyage historically possible. Da Gama did not know those waters. The pilot did. He knew the monsoon winds, the ocean currents, the star positions that allowed a 23-day crossing from Malindi to the Malabar coast. Without him, the expedition does not reach India. His name does not appear in the Portuguese record. Portuguese sources note only that a pilot was provided by the Sultan of Malindi, that da Gama's men assumed he was Christian without apparently establishing his name, and that he may have been taken back to Lisbon: his knowledge extracted, his identity unrecorded. What was received was used. What was used was not credited. What was not credited was eventually absorbed into the story of European discovery. This is not a claim of deliberate suppression. It is a description of how Portugal's recording system worked: it had a category for Portuguese achievement and no category for Indian contribution. The knowledge entered the system. The source did not. The calculus transmission debate asks whether a similar pattern operated in the mathematical domain a few decades later, involving Jesuit missionaries in the same Kerala, encountering the same tradition, returning to the same Europe. The navigator case does not prove that mathematical transmission occurred. It establishes that, in the same broad geography and within the same early colonial contact zone, Indian knowledge could be used by

Portuguese actors without preserving the identity of its source. This establishes a pattern of record, not a proof of transfer. The absence of a receipt, in that context, cannot be treated as decisive either way. It cannot establish transmission; it only prevents absence from being treated as proof of non-occurrence.

The point is not that this pattern proves transmission in the Kerala calculus case. It is that the absence of attribution, in a system that repeatedly failed to preserve it even when the knowledge transfer is undisputed, cannot be treated as decisive evidence against it. The absence of a receipt must be interpreted in light of how such receipts were — or were not — produced in comparable cases.

Where That Leaves Us

The Kerala case occupies the precise middle of the evidentiary spectrum. It is not pure equivalence: there is real, historically grounded circumstantial evidence of contact, motivation, and temporal proximity. But it is not documented transmission either: the record that would confirm it has not been found.

What the evidence supports is a carefully bounded set of possibilities. Transmission may have occurred through channels that left no surviving documentary trace: oral communication, informal exchange, knowledge that crossed the gap without leaving the kind of written record we need to confirm it. It may not have occurred at all, the observed convergence explained by independent development under shared intellectual pressures. The third possibility (stimulus diffusion, introduced in Chapter Three) also cannot be excluded. The shared need to improve astronomical tables for navigation and calendrical accuracy bore down on Kerala astronomers and European mathematicians simultaneously. That shared pressure could independently drive the development of increasingly precise trigonometric methods in both places, without any specific text or person crossing the gap.

None of these possibilities can be established from the evidence currently available. The honest position is to hold all three open and resist resolving the question before the evidence warrants it.

Keeping It Open

Keeping the question open is not a failure of resolution. It is the correct application of the standard this book has been building toward.

Apply the same standard in all directions. Where evidence meets it, affirm it; where it does not, hold the question open. In the Kerala case, that yields a precise and honest result: the priority of the results is established; the opportunity for transmission existed; the motivation was present; the documentary confirmation is absent. The question therefore remains open.

♦

I began this book by following a thread. A verse about triangular numbers led to Bhāskara II's pedagogy, which led to Āryabhaṭa's freedom to compose and transmit without fear, which led to the question of what evidence actually looks like when transmission is established or merely assumed. The Kerala case is where that thread runs into genuine darkness, not the darkness of ignorance, but the darkness of an incomplete record that has resisted the efforts of serious scholars for decades.

I have been sitting with this uncertainty for longer than any other question in the book. The Kerala School produced results of such quality that the question of whether they influenced Europe is genuinely interesting, which is itself a measure of how significant the results were. Mādhava's infinite series are not the less remarkable for being uncertain in their influence. What the thread led to here is not resolution but precision: we know what was achieved, and we do not know whether it crossed the gap. Uncertainty about transmission is not uncertainty about achievement.

What has been traced so far is how knowledge moved, how it was preserved, and how we evaluate claims about its transmission. But that leaves a more basic question unanswered: how is such knowledge generated in the first place, and what allows it to hold together with enough precision to survive across centuries?

To answer that, we have to look inside it. Not at the claims made about it, but at how it works: how observation is organized and how deviation is handled. What the next two chapters show is not decoration or digression. It is evidence: evidence that this tradition was a precision engineering enterprise, not a philosophical one, capable of sustaining extraordinary results for thirteen centuries. The mathematics is the proof of work.

Chapter Six

From Mean to Motion

How Āryabhaṭa Made the Heavens Computable

The Kerala School's infinite series — the achievement whose journey to Europe the last chapter could not confirm — did not arise from nowhere. It arose from a tradition of mathematical astronomy whose most important early architect was Āryabhaṭa. In Chapter Two he appeared as a young man composing and transmitting without fear. What made that act remarkable was not merely the cultural conditions that permitted it but the substance of what he had built: a computational system for the heavens — one that would be used across Asia for a thousand years to calculate calendars, predict eclipses, and determine planetary positions.

What follows is how he built it. The story begins not with the sky but with arithmetic — with the concept of the mean. And what Āryabhaṭa did with it is one of the most elegant moves in the history of mathematical astronomy.

The Problem That Would Not Stay Solved

Āryabhaṭa's problem was ancient and stubborn: planetary motion does not behave uniformly. Planets speed up, slow down, and sometimes appear to reverse direction entirely. Every astronomer before him had faced the same sky and the same embarrassment, the gap between any clean prediction and what the heavens actually delivered.

His first move was not to close that gap. It was to define it.

But before he could define it, he needed a concept sturdy enough to anchor it. That concept was the mean, and we have already seen it in this book. In Chapter One, tracing the lineage of the triangular numbers verse, the arithmetic principle appeared: the sum of a sequence equals the mean multiplied by the number of terms. Āryabhaṭa had worked with this not merely as a middle value but as a quantity that stands in for an entire sequence when properly scaled. That same instinct, isolate the mean and make it the anchor, is what he brought to the sky.

He assumed uniform motion, a mean position computed from time alone. In this framework, a planet follows a mean path, the path it would take if it moved uniformly. That mean path can be calculated for any moment in time using simple arithmetic. By committing to this fiction, Āryabhaṭa created a stable baseline against which reality could be measured.

The mean is not the answer. It is the instrument that makes the answer visible.

The Shape of the Error

Once the mean is in place, something unexpected happens. The deviation — the difference between the predicted mean position and the observed actual position — stops looking like chaos. It rises, reaches a maximum, falls, crosses zero, goes negative, finds a floor, and repeats. Smoothly. Periodically. With a patience that chaos does not possess.

What he recognized is that the error had a shape. Not noise, not accumulated error, but something structured and geometrically familiar. He had seen it before, not in the sky, but in the geometry of the circle.

Take a circle. For a given arc, draw the chord connecting its endpoints and take half of it. This half-chord is the *jyā*, the quantity that later mathematics would call the sine. As the arc grows from zero, the *jyā* starts small, rises steadily, peaks when the arc reaches a right angle, and falls again. It is a smooth, symmetric, self-returning curve. The geometry breathes in and out.

The discovery is not the sine table or the formula, but the recognition that planetary deviation and the behavior of the half-chord were the same, that the sky's irregularity and the circle's geometry were governed by the same structure.

As someone trained in genetics, I recognize the structure of this move immediately. In a chi-square test, you compare observed values against expected ones, and it is the pattern of the residuals, not the raw observations, that tells you whether something real is happening. A random scatter says noise. A systematic deviation says signal. Āryabhaṭa performed the astronomical equivalent fifteen centuries before the statistics were formalized: subtract what you understand, examine what remains, and refuse to stop until the residual gives up its shape.

The Architecture of the Table

Picture a large circle with a smaller circle whose center rides along its rim: the smaller turning as it travels, the planet carried on its edge. The mean position marks the traveling center of the smaller circle, advancing uniformly along the larger one, while the planet rides the rim of the smaller one.

The planet's actual position therefore does not coincide with its mean. It is displaced from it, sometimes ahead, sometimes behind — by an amount that varies smoothly as the planet moves around the smaller circle. That displacement is not arbitrary: it is governed entirely by the geometry. It grows from zero, reaches a maximum, falls back to zero, and reverses: exactly the pattern of the *jyā*.

The correction does not have to be imposed from outside. It emerges from the geometry of the construction itself. (If at any point the procedure here feels compressed, I've reconstructed it step by step — both visually and numerically, with worked examples — at phanikurada.com/aryabhata.)

This is the epicycle, the same geometric concept that appeared in Chapter Three as a point of methodological debate. There, the question was whether the structural similarity between Ptolemy's epicycles and Āryabhaṭa's corrections proved that one borrowed from the other. Seeing the construction from the inside, as Āryabhaṭa would have seen it, the question looks different. The epicycle is not a decoration or a borrowed tool. It is the natural geometric consequence of representing a periodic deviation using circular motion. Any tradition working on the same problem and reaching for circular geometry would arrive at something like it. The similarity is real; the inference of borrowing remains unproven.

But knowing the shape of the epicycle is only half the problem. The epicycle tells you that the correction grows and shrinks as the planet moves around its small circle, but to produce a number, you need the *jyā* for the specific angle the planet has reached at that moment. The geometry draws the curve, but the *jyā* table reads off the height of that curve at each point along it. With it, you can find the planet's true position at any moment from three quantities: where it started, how far it has traveled, and what the *jyā* says about the deviation at that angle. The table makes the epicycle usable.

With the construction in place, the sine table's design becomes fully intelligible. Twenty-four values, each stepping 3°45' around the quadrant, not an arbitrary count, but a considered one. Fine-grained enough for astronomical correction, yet compact enough to preserve and transmit in a culture where transmission meant memory. Instead of storing a lifetime of observed deviations, Āryabhaṭa stored the shape that produced them.

A quadrant (one quarter of the circle) is sufficient because the sine's symmetry means all four quarters can be reconstructed from one. Twenty-four values cover one quadrant, and the circle does the rest. Āryabhaṭa was not storing a sample. He was storing the minimum representation of a complete mathematical function.

Everything above collapses into a single line:

$$\textbf{True position} = \textbf{Mean} \pm \textbf{k} \cdot \textbf{\textit{jyā}}(\theta)$$

The only quantity that could not be derived from the geometry alone is *k*: the amplitude, the maximum extent of each planet's wandering from its mean path. That number is different for every planet and has to be won from observation. But once measured, it does not change. Set it once, and the formula works for that planet at any moment, forever. Everything the system does ultimately reduces to this.

Three quantities. Any moment in time. A computable sky.

Bhāskara I, writing his commentary on the *Āryabhaṭīya* in the 7th century, took this foundation further in a way that deserves notice. He developed a rational approximation for the sine function, a formula using only basic arithmetic that reproduces the *jyā* values to within about 1% across the full range of the quadrant. It is remarkable not only for its accuracy but for its opacity: scholars have verified that the formula works without being able to reconstruct how Bhāskara I derived it. This shows how far the tradition pushed computation within its own framework.

The Test: Predicting an Eclipse

A computational system is only as good as what it can predict. And nothing tests a model of the sky more severely than predicting an eclipse.

The Moon does not shine. It is a dark body, lit by the Sun, visible only because it reflects borrowed light. Āryabhaṭa did not simply note this as an ancient idea: he made it load-bearing inside a predictive model. If the Moon carries no light of its own, then when the Earth moves between the Sun and the Moon, the Moon does not merely dim. It loses its source entirely. It enters shadow. The same logic runs in the other direction: the Moon between Earth and Sun, its shadow falling on us. In both cases, the event is nothing more than an alignment, a predictable geometrical consequence of three bodies moving in calculable paths. And alignments, once you can compute positions, can be predicted.

This is where the full weight of the mean-and-correction model lands. Predicting an eclipse requires knowing not just approximately where the Sun and Moon are, but their precise positions, including the Moon's deviation from its mean path, corrected by exactly the *jyā*-based method Āryabhaṭa had constructed. The shadow falls when and only when the geometry demands it. The model either delivers that precision or it does not.

It did.

♦

The *jyā* table delivered that precision. But it did something beyond calculation as well. It demonstrated that the sky had a geometry, that its apparent chaos concealed structure, and that a patient mind armed with the right framework could strip away the supernatural and find something more astonishing underneath.

In Chapter Two, Āryabhaṭa set down without fear that eclipses are shadows, not the work of Rahu. Now we see what that claim actually was: not philosophy, not defiance, but a computational commitment. The shadow falls on schedule, at the moment the calculation says it will, indifferent to tradition, available to anyone willing to look and compute.

What strikes me about the *jyā* table is not its elegance but what preceded it: the decision to subtract the mean and examine what remained. That decision — so fundamental to scientific practice that it is rarely named explicitly — Āryabhaṭa built into the architecture of his system in the fifth century.

The computational foundation is now in place. But it leaves a question untouched: what does this system say about what is physically real? Is the earth stationary at the center of a revolving cosmos, or does it move? The *jyā* table works either way. The eclipses can be predicted. The numbers do not force a choice between the two pictures.

Āryabhaṭa made a choice anyway. And how he arrived at that choice — what kind of reasoning led him to assign the daily motion of the entire sky to the rotation of the ground beneath his feet — is the question the next chapter takes up.

The computation was complete. The deeper question was just beginning.

Chapter Seven

The Choice the Numbers Did Not Force

How Āryabhaṭa Arrived at the Rotation of the Earth

The previous chapter ended with a question the computational system could not answer. The *jyā* table works whether the earth moves or not. The mean-and-correction framework is indifferent to the question of what is physically real. It asks only: do the numbers work?

But Āryabhaṭa went further. And what he did next was not analogy. It was architecture. Alongside the revolutions of the Sun, Moon, and planets, he assigned the earth a precise number of rotations over a *Mahāyuga*, a vast astronomical cycle of 4.32 million years. He built a geometric scheme anchored at a theoretical point on the equator that works naturally with a rotating earth. He separated the earth's rotations from the solar day, treating them as distinct cycles. These are not philosophical asides. They are parameters, quantities meant to be used in calculation.

A system that did not assume the earth rotates would have little reason to track those parameters in this way. The question this chapter asks is: what led him there? Not what he claimed, that is clear. But how he arrived at it, and how seriously he meant it, and what it tells us about how a mind of the first order operates at the edge of what is knowable.

Beyond the Analogy

The most cited passage comes from the Gola section of the *Āryabhaṭīya*, the chapter on the celestial sphere. It does not declare "the Earth rotates" in a single direct sentence. Āryabhaṭa was not writing a manifesto. He was writing a Siddhānta. What the verses say — and what they imply — are two carefully distinct things.

अनुलोमगतिनौस्थः पश्यत्यचलं विलोमगं यद्वत् ।

अचलानि भानि तद्वत् समपश्चिमगानि लङ्कायाम् ।। ९ ।।

anulomagatinausthaḥ paśyaty acalaṃ vilomagaṃ yadvat |

acalāni bhāni tadvat samapaścimagāni laṅkāyām ||

"As one in a boat moving forward sees what is stationary moving in the reverse direction, so at Laṅkā the stationary stars are seen moving exactly westward."

उदयास्तमयनिमित्तं नित्यं प्रवहेण वायुना क्षिप्ताः ।

लङ्कासमपश्चिमगो भपञ्जरः सग्रहो भ्रमति ।। १० ।।

udayāstamayanimittaṃ nityaṃ pravaheṇa vāyunā kṣiptāḥ |

laṅkāsamapaścimago bhapañjaraḥ sagraho bhramati ||

"Driven constantly by the carrying wind (pravaha) for the sake of their rising and setting, the cage of stars together with the planets revolves exactly westward at Laṅkā."

– *Āryabhaṭīya, Golapāda* 9–10. Translated by the author from the Sanskrit.

Here "carrying wind" renders *pravaha*, literally "the streaming," from the Sanskrit root meaning to carry or bear. In traditional Indian cosmology, the *pravaha* is the cosmic current that carries the celestial bodies in their daily westward circuit, not a violent storm but a steady, carrying stream. The asterisms are the nakṣatras, the fixed star groupings that form the backdrop of the sky, distinct from the planets that wander through them.

Notice what these two adjacent verses together actually say. Verse 9 does not say the earth rotates: it says that stationary stars appear to move westward and offers a relativity analogy to explain why. Verse 10 immediately follows with the conventional cosmological picture: a stationary Earth, the sky rotating around it, the asterisms driven westward in their daily circuit by the *pravaha*, the provector wind. Verse 9 quietly reverses this picture; Verse 10 restates it. Both sit side by side, unreconciled. This is not confusion. It is a precise record of what Āryabhaṭa was doing: embedding a new inference within an inherited framework without demanding that the framework be rebuilt around it.

That the argument is inferential does not mean it was ambiguous. Brahmagupta, writing a generation later, understood the implication exactly, and spent considerable effort refuting it on physical grounds. You do not marshal arguments about birds being swept westward and stones falling sideways against a mere analogy. You marshal them against a claim you have understood clearly and find physically indefensible. Brahmagupta's fierce objection is itself the strongest evidence that Āryabhaṭa's contemporaries knew precisely what he meant.

Brahmagupta was not alone. Varāhamihira, writing earlier in the same century (d. 587 CE), had also attacked the rotation claim on physical grounds. Two serious astronomers, within the same intellectual tradition, in the same century, treating Āryabhaṭa's statement as a committed physical assertion worth sustained rebuttal, not a pedagogical device, not a relative-motion analogy, not a poetic flourish. Shukla and Sarma note in their critical edition that Āryabhaṭa's theory "received a severe blow at the hands of Varāhamihira (d. A.D. 587) and Brahmagupta (A.D. 628)," and that their arguments "could not be refuted by any Indian astronomer," not because Āryabhaṭa was wrong, but because the physics required to answer them did not exist yet.

There is a structural argument here that deserves to be stated directly. The *Āryabhaṭīya* is organized with deliberate functional precision: the *Gaṇitapāda* gives mathematical tools, the *Kālakriyāpāda* gives computational procedures for time and planetary motion, and the

Golapāda gives the physical model of what is actually happening in the sky. In a text this compressed (121 verses covering all of mathematical astronomy) nothing is placed carelessly. Earth's rotation is not in the computational sections. It is in the model section. That placement is itself a statement of intent: this is not a computational parameter but a physical interpretation, placed exactly where physical interpretations belong. In a sutra-style text, inclusion signals importance; placement signals function.

The computation is frame-neutral: the sine tables, the epicycles, the corrections all work equally under a rotating sky or a rotating earth. But Āryabhaṭa does not leave the frame entirely unspecified. He places rotation in the model section, strongly suggesting he intended it as a physical commitment rather than a purely computational device. He gives you the machinery and separately tells you what he thinks is actually moving. The separation is not evasion. It is architectural elegance: preserve computational continuity, state physical commitment, leave the reader to hold both.

The Language of Parameters

The *Āryabhaṭīya* is not a philosophical treatise. It is a dense computational manual, economical and focused on results, organized around cycles and counts. In this tradition, mathematical models were valued as instruments for reliable calculation, not necessarily as literal descriptions of the physical world.

The computational framework (the *jyā* table, the mean-and-correction formula, the eclipse prediction method) was designed to produce accurate predictions regardless of which physical picture was correct. What the system was built to answer was not 'what is physically real' but 'where will the planet be.' Individual thinkers like Āryabhaṭa made physical commitments: this chapter has examined that carefully, but the computational architecture did not require them. A practitioner could use the system without resolving the physical question. That separation is precisely what made the system robust across a millennium, and precisely what meant that a structurally significant result like Nīlakaṇṭha's — the Kerala astronomer whose 1501 reorganization of the inner planets around the Sun is examined in Chapter Nine — could be absorbed as a computational refinement without demanding a physical reckoning.

Within that framework, one detail stands out. Āryabhaṭa distinguishes between the earth's rotations and the solar day itself. In his system, the number of earth's rotations over a *Mahāyuga* is slightly greater than the number of solar days, implying that a single rotation of the earth is not identical to a day as measured by the Sun. This separates the earth's true rotation from the apparent daily motion of the Sun, treating them as distinct cycles rather than the same phenomenon.

This is parameter language. It is something meant to be used in calculation. A system that did not assume the earth rotates would have no reason to track its rotations as a quantity distinct from the solar day. The fact that Āryabhaṭa does this — and builds his timekeeping around it, using the stellar cycle as the primary reference rather than the solar one — strongly suggests a framework built to accommodate that assumption rather than a system into which the rotating earth has been inserted as an afterthought.

The Tension He Left in the Text

The picture is not clean. And the fact that it is not clean is itself informative.

Verse 10 restates the conventional picture even as Verse 9 quietly departs from it. He does not reconcile these two descriptions. Both remain in the text, sitting alongside each other without resolution.

Āryabhaṭa was not trying to dismantle one worldview and replace it with another. He was working within an inherited framework while quietly reassigning where the motion lives, leaving the older language in place even as his mathematics quietly departed from it.

This is a recognizable pattern in the history of science. Copernicus kept much of the epicyclic mathematical apparatus of Ptolemaic astronomy even while shifting the center from Earth to Sun. A thinker at the frontier of what is knowable often works within inherited structures while pushing against their limits, not because they are confused, but because they cannot yet see everything that would need to change if their insight were fully correct. Āryabhaṭa reassigned the motion without explaining all that would follow from the reassignment. The *pravaha* stayed in the text. The tension stayed unresolved. And the system worked.

Why Rotation

From a modern standpoint, Āryabhaṭa did not prove that the earth rotates. The same observations can be explained equally well by a rotating sky. Nothing in his work compels one interpretation over the other.

But his choice was not arbitrary either.

He appears to have recognized that the daily motion of the heavens admits two explanations: either the sky turns, or the observer does. Later discussions would describe the second possibility as *bhūbhramaṇa*, the rotation of the earth. Having seen both possibilities, Āryabhaṭa assigned the motion to the earth and built his system accordingly.

The reason is one that Chapter Six has already prepared the reader to understand. A rotating sky requires imagining the entire celestial sphere in perpetual daily motion. A rotating earth allows that motion to be localized to the observer. For a system built on cycles, counts, and coordinates, for a computational manual whose purpose was accurate prediction, the second arrangement is cleaner. It does not make it true. But it makes it useful. And in a Siddhānta, usefulness is the operative criterion.

The same instinct that produced the *jyā* table and the eclipse prediction led him here too. Localize the motion. Build the system around the more economical description. Trust the mathematics over the intuition.

The Objection That Could Not Yet Be Answered

Even without formal physics, an obvious difficulty presents itself: if the earth were spinning, why are objects not flung away from its surface? Why does a dropped stone fall straight down rather than being carried sideways? And: most vividly, if the earth is rotating rapidly eastward, why do birds that take flight not get left behind, swept westward while the ground turns beneath them?

Brahmagupta, writing in the seventh century, makes these objections explicit. His rejection of earth's rotation is not obscurantism and not the defense of religious cosmology. It is a reasonable demand that a model explain behavior within the world, not just appearances above it. Without the concept of inertia — the idea that objects in motion tend to stay in motion, and that a spinning earth carries its surface and atmosphere with it — a rotating earth has no convincing physical basis. Brahmagupta was asking exactly the right question. He arrived at a conclusion that later physics would not support, but the question itself was correct.

What makes his position more interesting than mere objection is what he did have. In the same *Brāhmasphuṭasiddhānta* in which he argues against rotation, Brahmagupta names and describes *gurutvākarṣaṇam*, a natural tendency of bodies to fall toward the earth. He states that it is in the nature of the earth to attract and keep things, just as it is in the nature of water to flow. This is not a casual observation. It is a named concept, articulated in a technical astronomical text, identifying a real physical regularity: more than a millennium before Newton, and in a form more explicit than the natural-motion explanations found in other traditions at the time.

What he could not yet do was connect this insight to the problem he was confronting. Attraction is why objects fall; it does not explain why they are not left behind by a moving earth. The missing piece was inertia. He had one part of the framework. The part that would connect attraction to rotation was still a millennium away.

There is a deeper irony here. The gurutvākarṣaṇam passage sits in the same text where Brahmagupta uses gravity as a closing argument against Āryabhaṭa: earth attracts things downward, therefore a rotating earth would fling them sideways, therefore Āryabhaṭa is wrong. Brahmagupta was not arguing from obscurantism. He was head of the Ujjain observatory, responsible for the computational system on which calendars, eclipse predictions, and ritual timing depended. The rotating earth threatened that system's physical coherence — not the accuracy of its predictions, since the mean-and-correction framework produces correct eclipse timings whether the earth rotates or the sky does, but the physical picture the model claimed to describe. Gravity gave him the tool to close the threat. The insight was genuine: a named physical regularity, but it was recruited immediately into the service of institutional defense. Once the debate was settled, the question closed. No one asked what else that regularity might explain. The tradition did not suppress it. The tradition found it useful before it could become generative.

That is absorptive pressure at the level of physics: a genuine insight recruited into institutional defense before it could generate a question in its own right.

Brahmagupta had, in the same text, named the phenomenon that would eventually make Āryabhaṭa's model physically intelligible, and used it as a reason to reject it.

Āryabhaṭa, facing the same observations and the same implicit objection, allowed the model to stand anyway. He did not explain how a rotating earth could leave ordinary experience undisturbed, that explanation would not become available until Galileo and Newton, more than a millennium later. What he did instead was permit a coherent mathematical system to exist even when it conflicted with the physical intuition of his time.

The contrast between Āryabhaṭa and Brahmagupta is not who was right, but which pressure each yielded to. Brahmagupta yielded to the pressure of physical common sense: a reasonable pressure, and one that most careful thinkers in the absence of inertia would have yielded to. Āryabhaṭa yielded to the pressure of mathematical coherence. The system worked better with the earth rotating. He trusted the mathematics over the intuition.

This was the tension the argument had been circling from the beginning: a system that works, and a question it does not force.

That trust was not vindicated in his own lifetime, or in Brahmagupta's, or for centuries after both of them. It would require inertia, the telescope, and the unification of terrestrial and celestial mechanics to fully justify what Āryabhaṭa's mathematics had quietly assumed.

What Kind of Discovery This Was

Experimental verification is not the right measure for what Āryabhaṭa did, not because the standard is too demanding, but because it asks the wrong question of what he was doing. He was not reporting an observation. He was choosing between two descriptions that fit the same observations equally well.

His contribution is specific: he recognized that observed motion does not uniquely belong to the heavens, saw that assigning it to the earth was equally consistent with the observations, and built an astronomical system in which that assignment works naturally and produces reliable results. Whether he believed in the physical reality of earth's rotation, or treated it as the most computationally elegant interpretation, is a question the text genuinely cannot answer. The Siddhāntic tradition itself does not always sharply separate calculation from physical description, which makes that distinction difficult to draw with confidence.

What is clear is this: a mind was willing to separate appearance from underlying description, to embed an unfamiliar possibility into a working system before anyone could rigorously justify it, and to leave the tension unresolved rather than forcing a false coherence. Later astronomers in his lineage would take that unresolved tension seriously, proposing physical accounts to reconcile the motion of the earth with the behavior of objects within it. They were working on the problem Āryabhaṭa had bequeathed them.

Āryabhaṭa did not know that. He was simply trying to make the numbers work.

♦

That is the achievement Chapters Six and Seven have been tracing together: a tradition that built a computational system precise enough to predict eclipses, flexible enough to accommodate a rotating earth, and honest enough to leave its unresolved tensions in the text rather than hiding them. The *jyā* table, the mean-and-correction framework, the separation of stellar and solar cycles, the rotating earth embedded as a parameter rather than a claim: these are the products of minds working with exceptional care on a genuinely hard problem.

The system was remarkable. It was also genuinely limited, not by the quality of the minds working within it, but by the question those minds were asking. The limitation was not in the mathematics. It was in the question.

That question would eventually be asked, but the path to it ran through a thousand more years of development, and through a narrow coastal strip in the far south of India that would push Indian mathematical astronomy further than anyone standing in Āryabhaṭa's world could have imagined. The next chapter traces that path: what Kerala is, why it became what it became, and what the tradition built there before Nīlakaṇṭha Somayājī sat down to write the *Tantrasaṅgraha*.

Chapter Eight

The School at the Edge of the World

Haridatta, Mādhava, Parameśvara, and What Isolation Made Possible

The question has a geographic form before it has a mathematical one.

Kerala is not where you would place a great astronomical school if you were designing one. It is a narrow coastal strip, bounded on the east by the Western Ghats and on the west by the Arabian Sea, a strip roughly 600 kilometers (370 miles) long and never more than 120 kilometers (75 miles) wide. Kerala was not isolated from the world: it was one of the great maritime trading zones of the ancient and medieval world, a destination for Arab merchants, Chinese ships, and Roman spice buyers long before Vasco da Gama arrived at Calicut in 1498. But trade routes and scholarly networks are not the same thing. The intellectual networks of the Nambudiri Brahmin communities ran inward, through the river valleys and temple traditions of the interior. Ujjain, the meridian city around which Indian astronomy had organized itself since before Āryabhaṭa, sat 1,500 kilometers (930 miles) away across terrain that the Ghats made genuinely difficult to cross.

What that geography made was not limitation. It was ownership: of a tradition, its failures, and the obligation to solve problems no one else could be asked to solve.

And yet, between roughly the seventh century CE and the sixteenth, one of the most mathematically sophisticated astronomical traditions in the world operated in this narrow coastal strip. It computed π to seventeen decimal places. It developed the infinite series for sine and cosine — a method for computing exact trigonometric values through an unending sequence of shrinking terms that converges on a precise answer — which Gregory and Leibniz would independently discover in the 1670s. It produced, in the *Yuktibhāṣā* of Jyeṣṭhadeva, the first systematic statement of the proofs underlying calculus-like methods:

written in a vernacular language, in a form designed to be followed rather than merely memorized.

The forensic question is why. Not what it produced, but what sequence of pressures and conditions produced it. The answer, it turns out, is inseparable from the geography. The scholarly isolation that made the tradition invisible to the wider Indian astronomical world was the same isolation that drove its mathematics so deep.

The Kerala school was not an independent invention. It was the Aryabhatan tradition, localized by Haridatta in 683 CE and deepened over eight centuries without ever departing from that foundational commitment. Carried south through the scholarly networks of Brahmin communities, adopted as a working framework, used for generations: what changed in Kerala was not the origin of the system but its ownership. Once Haridatta adapted the parameters to the local meridian, the tradition was no longer something Kerala consulted; it was something Kerala maintained. A borrowed system can always be returned to its source when it fails. An owned system cannot. Kerala had no Ujjain to appeal to. It had only itself.

The Meridian Problem

A meridian is a line of longitude, an imaginary north-south line running through a fixed location on the earth's surface. In astronomy it serves as a reference frame: all planetary positions, all eclipse timings, all calendrical calculations are computed relative to a specific meridian, then adjusted for wherever the practitioner actually stands. Get the reference meridian wrong, or fail to correct for it accurately, and every calculation built on it carries the error forward. The story begins not with mathematics but with a practical failure of exactly this kind, in the seventh century CE, roughly 200 years after Āryabhaṭa.

The astronomical tradition Āryabhaṭa had built was organized around Ujjain. His parameters (the cycles, the counts, the reference epochs) were calibrated to the Ujjain meridian, the longitudinal reference point that Indian astronomy had used since antiquity. A practitioner using the *Āryabhaṭīya* computed planetary positions relative to Ujjain, then applied a correction for the difference in longitude between Ujjain and wherever they actually were. This was standard procedure. It worked.

It worked, that is, if the longitude correction was accurate. For practitioners in Kerala – at the southwest tip of the subcontinent, roughly 1,500 kilometers (930 miles) southwest of Ujjain – the correction was not trivial. Ujjain sits near the geographic center of India; Kerala sits at its far southwestern edge. When the Ujjain-calibrated system said an eclipse would begin at a given moment, Kerala practitioners found it beginning minutes earlier or later than predicted, a gap that the standard longitude correction never quite closed. For calendrical purposes, this was tolerable. For eclipse prediction, it was not.

Into this situation, around 683 CE, came a scholar named Haridatta: a Nambudiri Brahmin astronomer working in Kerala, trained in the Aryabhatan tradition and serving the ritual and calendrical needs of the communities around him. What he did appears, on the surface, modest: he adapted Āryabhaṭa's parameters to the Kerala meridian rather than Ujjain's. The result was the Parahita system: para meaning suitable for others, hita meaning beneficial. A system calibrated to where you actually were.

The name is worth pausing on. It does not announce itself as a correction to Āryabhaṭa, or an improvement on the standard tradition, or a new departure. It announces itself as something made for use: locally useful, locally owned. Haridatta was not trying to reform Indian astronomy. He was trying to make it work for the communities he served.

But the consequence of that localization was something he could not have fully anticipated. The fix would have to come from within.

Fifty-Five Years of Disagreement

The failure, when it came, was quiet. It did not announce itself as crisis. It accumulated.

By the fourteenth century, the Parahita system had been in use for roughly 700 years. Parameters that had been reasonable approximations in Haridatta's time had drifted against the actual sky. The moon's position at eclipse was not quite where the computation placed it. Planetary conjunctions fell slightly early or slightly late. The errors were not catastrophic: the system still produced usable results for calendrical purposes. But they were real, and they were systematic. A practitioner careful enough and patient enough to watch the sky over a long period would see them.

Parameśvara was that practitioner.

Parameśvara was born into a Nambudiri Brahmin family in Vatasseri, in what is now the Malappuram district of Kerala, a family with a tradition of astronomical scholarship reaching back generations. He began systematic observational records around 1393 CE and continued them for roughly fifty-five years. He was not observing casually. He was comparing: computed position against observed position, prediction against event, the numbers the Parahita system gave him against what his instruments and his eyes actually found.

What he found, over those fifty-five years, was that the disagreement was not random. It was systematic: the same errors appearing in the same calculations, stubborn across parameter adjustments that should have fixed them.

His response was the Dṛggaṇita system: dṛg meaning eye, gaṇita meaning computation. Eye-computation. The name is a methodological statement: when the eye and the calculation disagree, the eye is the authority. Trust the sky over the inherited numbers.

This sounds obvious. It was not. A tradition that had carried Āryabhaṭa's parameters for nearly a millennium, through commentaries and refinements and successive generations of scholars, had an enormous investment in those parameters. To say the sky overrides the tradition was not a small claim. It was a reordering of what counted as evidence.

His primary instruments were the gnomon, the armillary sphere, and a water clock. The work was systematic, documented, and sustained over half a century. A well-made gnomon and a carefully calibrated armillary sphere could detect errors on the scale of several arcminutes (one arcminute being one sixtieth of a degree): large enough to show that the Parahita parameters were drifting, large enough to demand a response. What they could not do was detect the one- or two-arcminute errors that would make a small residual undeniable. That precision would arrive a century later, with Tycho Brahe's fixed observatory at Uraniborg. The gap was not in intelligence or dedication but in physical infrastructure, in what the instruments could force into the open. That gap is why the step Kerala approached was left for Europe to take.

Parameśvara was not the first Indian astronomer to make observations. The tradition had always intersected with practice. What distinguished him was scale and intention: fifty-five years of records, systematically compared, treated as a verdict rather than a check. He was not verifying the tradition. He was testing it. And in testing it, he identified something specific. The errors in the Parahita system were not fixable by adjusting the existing parameters within the existing mathematical framework. They were structural. They lived in the mathematics itself, in the method of computing sines.

The Interpolation Problem

Āryabhaṭa's table gave the half-chord, the sine, in modern terms: for 24 angles, spaced 3.75 degrees apart, from 3.75 degrees to 90 degrees. This was sufficient for most astronomical calculations. When you needed the sine of an angle not in the table, you interpolated, estimated the value between two known points based on how far between them your angle fell.

The problem is that the sine is not linear. Between any two table entries, it curves. Linear interpolation introduces an error: small at each step, but compounding across calculations.

To eliminate the interpolation error, you needed a way to compute the sine of any angle directly, not by estimating between known values, but by calculation from first principles, to whatever precision the situation demanded.

This was the precision problem that Mādhava's series resolved, not as a response to Parameśvara's specific observations, but as part of the same tradition's broader drive toward exactness that teacher and student were pursuing together.

The Machine Mādhava Built

Mādhava of Saṅgamagrāma lived from approximately 1340 to 1425 CE. He was Parameśvara's teacher: forty years older, already developing his mathematical work when Parameśvara began his systematic observations around 1393. The two were not in sequence: mathematics first, then observation, or observation first, then mathematics. They were parallel: a teacher working on the theoretical problem and a student working on the observational one, both pressing against the same precision ceiling from different directions. The series and the sky-watching were a conversation, not a relay.

What Mādhava discovered is now known as the infinite series for sine and cosine, and for π. In European mathematics, James Gregory reached the series for arctangent in 1671, and Leibniz and Newton reached the sine and cosine series around the same period. Mādhava was there approximately 250 years earlier. The mathematics is recognizably the same in structure, though reached through a different path.

Think of walking toward a wall by always covering half the remaining distance. You never technically arrive, but you get closer and closer, and the total distance converges on the wall. The sequence 1 minus

1/3 plus 1/5 minus 1/7 plus 1/9, continuing forever, converges on $\pi/4$. The alternating signs and odd denominators are not arbitrary: they trace back to the geometry of how arc length relates to chord length in the unit circle — the same *jyā* construction extended to the limit.

This is not approximation. It is a machine for exactness. The interpolation error disappears, because you no longer need to interpolate: you compute directly. Where Āryabhaṭa's table gave you a finite set of known values and left you to interpolate between them, Mādhava's series runs as far as you need it to, producing the sine of any angle, or the value of π, to whatever precision that need demands.

The extraordinary mathematics was the engineering response to a practical crisis.

Astronomical pressure and mathematical response reinforced each other.

Numbers Encoded in Language

Mādhava did not write his series as a formula in the modern sense. He encoded it as a verse, and the method of encoding is itself a piece of evidence about the tradition this book has been tracing.

The verse uses a system called *kaṭapayādi*, in which each consonant of the Sanskrit alphabet corresponds to a specific digit. A skilled reader can decode a seemingly ordinary verse: one that scans as poetry and carries some surface meaning, and extract from its consonants a precise numerical value. The verse simultaneously reads as language and encodes as number. Mādhava's verse for the circumference-to-diameter ratio gives, when decoded through this system, a value of π accurate to eleven decimal places: a specific result encoded in verse form; the series itself, extended further, reaches seventeen places.

The verse can be memorized. The verse can be transmitted across generations without writing. And anyone who knows the *kaṭapayādi* key can recover the numerical value at any time.

The verse is simultaneously a poem and a data structure. What changed between Bhāskara II's verse about triangular numbers and Mādhava's verse encoding π was not the principle. It was the depth of what needed to be carried. The container was the same. The contents had grown.

The Proofs Written Down

By the early sixteenth century, the Kerala School had accumulated several generations of results — Mādhava's series, Parameśvara's corrected parameters, Nīlakaṇṭha Somayājī's geo-heliocentric planetary model (the subject of Chapter Nine) — that had been transmitted primarily as versified results in the Aryabhatan tradition. You knew the result. You knew the procedure. The reasoning behind the result, the proof that it was true rather than merely useful, lived in the oral commentarial tradition, passed from teacher to student, demonstrated rather than written.

Around 1530, thirty years after the *Tantrasaṅgraha*, Jyeṣṭhadeva wrote the *Yuktibhāṣā*, a Malayalam exposition of the reasoning behind Nīlakaṇṭha's work. He belonged to the same Kerala lineage, by tradition a student of Dāmodara, but came a generation after Nīlakaṇṭha.

The text is unusual in several ways, each of which is informative. It is written in Malayalam (the vernacular language of Kerala) rather than Sanskrit, which had been the language of all serious astronomical and mathematical scholarship in the tradition. It is organized as a systematic exposition of proofs: not results, not procedures, but the reasoning that establishes why the results are correct. And it covers the infinite series results, the trigonometric series, and the foundations of what we would now recognize as calculus — integration through geometric reasoning, summation of infinitesimals, the relationship between rate and accumulation — approached without the modern formalism but with genuine logical rigor.

There is no other text like it in the Indian mathematical tradition.

The choice of Malayalam rather than Sanskrit requires interpretation. Sanskrit was the language that traveled: the learned language that scholars across the subcontinent shared, the medium of texts that were copied, commented on, and distributed through networks reaching from Kerala to Kashmir. A result written in Sanskrit could in principle reach Varanasi or Navadvipa. A result written in Malayalam was, at the level of language, already contained within Kerala.

Jyeṣṭhadeva was not trying to reach the wider world. He was trying to be understood.

This is recognizably the move Bhāskara I had made nine centuries earlier when he unpacked Āryabhaṭa's compressed sutras into patient procedural prose, the same instinct to make the mechanism visible rather than merely transmit the result. What changed is the scale of what needed unpacking. Bhāskara I was making one text accessible. Jyeṣṭhadeva was writing out the proofs that had underpinned a school's work for 150 years.

The *Yuktibhāṣā* is the Kerala School explaining itself to itself. The proofs it contains were almost certainly known orally before Jyeṣṭhadeva wrote them down. What the text represents is a moment of consolidation: a tradition deciding that the reasoning, not just the results, needed to be preserved in a form that could survive the loss of the teacher. That decision suggests a community that knew, on some level, what it had built.

♦

Between the seventh century and the sixteenth, a set of Nambudiri Brahmin communities on the southwest coast of India had taken an astronomical tradition that originally belonged to the Ganges plain, localized it to their own meridian, identified its failures through sustained and systematic observation, built the mathematical infrastructure to address those failures, and finally written out its proofs in the vernacular so that the reasoning, not just the results, could survive.

That sequence — localize, test, fail, build, prove — is the structure of a genuine scientific tradition. The mathematics was not the tradition's goal. It was the tradition's instrument.

What Kerala shows is not just that the tradition could produce deep mathematics, but how: when a system owns its own failure, it is forced to invent.

What Kerala had built was everything needed to go further, except the condition that makes further unavoidable.

Everything necessary for the next step was present. Nothing made it necessary.

Nīlakaṇṭha Somayājī was born around 1444 CE, in a village north of Thrissur. He studied under Dāmodara: Parameśvara's son, making him Parameśvara's intellectual grandson and Mādhava's intellectual great-grandson. He was not at the margins of this tradition. He was at its center, standing on three generations of accumulated precision. He inherited not a system to use, but a problem that had already been pushed to its edge. What he built with those tools — a geo-heliocentric model in which all five planets orbit the Sun — is the subject of the next chapter. He reached it independently, decades before Tycho Brahe arrived at the same position in Europe.

Chapter Nine

The Geo-Heliocentric Equilibrium

Nīlakaṇṭha Somayājī and the Geo-Heliocentric Model

The previous chapter established what Nīlakaṇṭha inherited: a tradition that had taken practical computation as its organizing principle for a millennium, producing extraordinary mathematical tools without directing them toward cosmological questions. What he did with those tools — and why his achievement belongs in the first rank of astronomical history — is the subject of this chapter.

The Tradition at Its Height

By the time Nīlakaṇṭha sat down to write the *Tantrasaṅgraha*, the Indian astronomical tradition had been developing for a thousand years. Āryabhaṭa's mean-and-correction framework had been refined by generations of astronomers. Brahmagupta had improved the constants and sharpened the corrections. In the centuries between Bhāskara II and Mādhava, mathematicians including Śrīpati and Śrīdhara continued the tradition's development, refining methods and extending results within the established framework. Mādhava then developed the infinite series that gave the tradition unprecedented mathematical precision. The tools, by the time they reached Nīlakaṇṭha, were extraordinary.

What it had built, in essence, was a system of remarkable computational power oriented toward a specific goal: accurate prediction of planetary positions for calendrical and ritual purposes. The *jyā* table, the mean-and-correction formula, the eclipse prediction method: all of these served that purpose with extraordinary effectiveness. The system worked. Generations of practitioners used it, trusted it, and refined it. Its success was not in question.

The Solar Organization Made Explicit

In the standard Indian planetary model, the motions of Mercury and Venus were explicitly tied to the Sun. Their corrections, the śīghra corrections (the swift epicyclic adjustments encoding each planet's relationship to the Sun), were keyed to the Sun in a way the outer planets' corrections were not. This was not accidental. It reflected a genuine observational truth: Mercury and Venus never stray far from the Sun in the sky because they orbit inside the Earth's orbit. Any model that reproduces their observed behavior accurately must, in some sense, encode their relationship to the Sun.

Indian astronomers encoded it correctly. They produced accurate predictions. But the encoding was computational rather than structural: the solar relationship was built into the correction terms without being treated as a statement about physical arrangement.

For eight centuries, recognition did not become reorganization.

Nīlakaṇṭha closed this gap. His planetary model was the most structurally original development in Indian astronomy since Āryabhaṭa. He took the implicit solar relationship that had been encoded in Mercury's and Venus's correction terms and made it explicit: in his model, Mercury and Venus orbit the Sun directly. The Sun itself then orbits the Earth, carrying the inner planets with it.

The model is known in the history of astronomy as a geo-heliocentric model: the Earth remains at the center of the cosmos, but the planets orbit the Sun rather than the Earth directly. The *Tantrasaṅgraha*'s specific achievement was for Mercury and Venus: it made their solar relationship structural rather than merely computational, and produced equations of center for those planets more accurate than any contemporary European model. But Nīlakaṇṭha did not stop there. In the Āryābhaṭīyabhāṣya (his extensive commentary on Āryabhaṭa's foundational text), he extended the geo-heliocentric model to all five planets. Mercury, Venus, Mars, Jupiter, and Saturn all orbit the Sun, which in turn orbits the Earth. This is structurally comparable to the model Tycho Brahe would develop independently in Europe roughly eighty years later — reached through a very different path, by following what the śīghra corrections implied about the physical structure of the solar system, without the competitive cosmological pressure that drove European astronomical development.

What drove him to that reorganization was not a search for a new cosmological architecture. It was a specific technical problem that had troubled Indian astronomers for eight centuries. The latitude calculation for Mercury and Venus — how far above or below the ecliptic these planets appear — was structured differently from the calculation for the outer planets, in a way that no one had satisfactorily explained. Bhāskara I noticed the asymmetry in the seventh century and recorded it without resolution. Around 860 CE, Pṛthūdakasvāmin returned to the same procedure and offered the only justification the tradition would settle on: it had been found, he wrote, to lead to predictions in conformity with observations. Three centuries later, Bhāskara II quoted that justification in his *Vāsanā-bhāṣya* and moved on. The anomaly was known, acknowledged, and unresolved across a tradition that was otherwise exceptionally precise. Nīlakaṇṭha's restructuring solved it. By anchoring Mercury and Venus in solar orbit rather than treating their solar relationship as a correction term, the latitude calculation for the inner planets became unified with the outer planets for the first time. The tradition received this as a solution to the latitude problem. It was. But it was also a structural statement about where Mercury and Venus belong in the solar system, a statement the tradition did not need to make to use the solution.

Picture the difference: before Nīlakaṇṭha, the solar relationship was embedded in Mercury and Venus's correction terms: present in the calculation, invisible in the architecture. After, it was explicit: these planets orbit the Sun. The reason was now in the model, not just the result.

The question is why the extension to all five planets required a separate work. The answer lies in what observational astronomy makes obvious and what it requires inference to reach. Mercury and Venus never stray far from the Sun in the sky. Their visual behavior announces their solar relationship to any careful observer. The śīghra corrections that encoded that relationship were, in this sense, catching up with something the sky had been showing for centuries. Making that relationship structural was, for the inner planets, a relatively conservative step: it solved a specific problem, clarified something already implicit, and remained within the tradition's computational orientation. Mars, Jupiter, and Saturn are different. They wander widely across the sky and retrograde in ways whose solar connection is not visually announced. Nothing in their observed behavior compels the conclusion that they orbit the Sun. Extending the solar organization to the outer planets required not observational compulsion but architectural generalization, following the structural logic of what had been done for the inner planets to its consistent conclusion. The Āryābhaṭīyabhāṣya, as a commentary rather than a computational manual, gave Nīlakaṇṭha the reflective space to do exactly that. Where the *Tantrasaṅgraha* was compact and result-focused, the commentary was interpretive and expansive. It was there, thinking through the implications of what he had already built, that he extended the principle to all five.

The remaining step — the step neither Nīlakaṇṭha nor Tycho Brahe would take — was not a further mathematical generalization. It was a cosmological commitment of a different order: removing the Earth from the center entirely. Asserting that the Earth moves. That the fixed point around which everything else revolves is not fixed at all. The mathematics required no new tools. The commitment required something the mathematics alone could not supply, a forcing condition that made the Earth's stillness impossible to maintain.

Nīlakaṇṭha completed the *Tantrasaṅgraha* in 1501. Copernicus began writing *De Revolutionibus* around 1513. These are not figures separated by a millennium: they are contemporaries, confronting the same structural problem in planetary astronomy, possessing broadly similar mathematical tools. Nīlakaṇṭha's latitude solution for Mercury and Venus was in some respects more accurate than anything in the Copernican tradition. The gap between them was not a gap of time, or talent, or mathematical sophistication. It was a gap of competitive context. Copernicus entered a tradition where cosmological frameworks were in active, unresolved competition — Ptolemaic, Copernican, and (within a generation) Tychonic — none of them reconcilable with the others. Each framework needed to show where the others failed. That competition created the pressure that made a small anomaly into a weapon. Nīlakaṇṭha was working in a tradition with no equivalent competition. Not because his institutions were less sophisticated, but because the specific historical collision that produces irreconcilable framework competition had not occurred there. The Ptolemaic-style cosmological framework was never entangled with religious doctrine in India to the same institutional extent. There was no Reformation fracturing authority and making cosmological allegiance a political act. The conditions that turned theoretical disagreement into framework war did not converge in Kerala. That convergence was not something any institution planned or demanded. It was a historical accident. And it is that accident, not any institutional inferiority, that explains the difference. Historical accidents are everywhere. What matters is whether a system is structured in a way that allows those accidents to become decisive: to tip a residual into a weapon, a disagreement into an irreconcilable collision. The structure does not cause the accident. It determines whether the accident matters.

What Nīlakaṇṭha had built was a complete structural model. The question this chapter now needs to address is not why he did not go further. It is what going further would have required: from anyone, anywhere, in any tradition.

What followed Nīlakaṇṭha's model is as revealing as the model itself. Later astronomers in the Kerala School adopted the geo-heliocentric model, refined its techniques, and improved its parameters. It became part of the tradition's standard toolkit. What neither

Nīlakaṇṭha himself nor any of his successors did was remove the Earth from the center. The final commitment — placing the Sun at the center and accepting that the Earth itself moves — was the step the tradition did not take. A working solution was received as a completion.

Geo-heliocentrism was not a failed heliocentrism. It was a historically stable solution under pre-Keplerian observational conditions: valid for its time, adopted correctly, refined appropriately, and eventually superseded not because it was wrong but because more precise observations made something better both possible and necessary.

Pṛthūdakasvāmin's earlier formulation, that the procedure had been found "to lead to predictions in conformity with observations," is worth pausing on. It does not say the procedure is correct. It does not say its structure has been understood. It says only that it works. The tradition had inspected the asymmetry, found no internal account for it, and chosen to leave it where it was. The discrepancy was articulated, transmitted forward, and explicitly designated as something that need not be resolved as long as the computation produced the right answer. The mechanism this book has been describing is named, here, from within the tradition itself, nine hundred years before Nīlakaṇṭha set out to undo it.

A local maximum is not only stable: it is, from within, invisible as a maximum. Every partial departure looks like deterioration. The only path forward requires reconceiving the framework entirely. The geo-heliocentric architecture had separated two things that a fully heliocentric model must unify: the Sun's computational centrality (organizer of all planetary motion) and the Earth's experiential centrality (coordinate origin from which the sky is read). In Kerala's system, both coexisted without forcing a contradiction. Nothing required them to become the same thing. A system without internal contradiction has no internal reason to change.

Kepler, a century later, would name this discrepancy himself, though he had no way of knowing about Nīlakaṇṭha. Writing about his European predecessor, he observed that Copernicus, ignorant of his own riches, took it upon himself for the most part to represent Ptolemy, not nature, to which he had nevertheless come the closest of all. The judgment was specific. Copernicus had inherited Ptolemy's procedure for the inner planets: applying their equation of center to the mean Sun rather than to the planet itself, and reproduced it inside the heliocentric framework without correcting it. He had the larger architecture. He carried the smaller error forward. Nīlakaṇṭha, working without the heliocentric framework, had already corrected what the heliocentric framework would still get wrong more than a century later.

There is a historical observation that sharpens this point. The physical size of the Sun, its vast superiority over the Earth in scale, was not unknown to these traditions. Aristarchus of Samos, working in the third century BCE, estimated the Sun to be roughly twenty times the diameter of the Earth. The actual ratio is closer to one hundred and nine. He made the argument explicitly: the smaller body should orbit the larger. He proposed heliocentrism on precisely that reasoning, in 270 BCE, roughly eight centuries before Āryabhaṭa. He was ignored for eighteen centuries, not by some distant civilization, but by his own. The Mediterranean intellectual lineage that ran from Ptolemy through Byzantium into medieval Europe had the heliocentric proposal in its archive the whole time, and did not act on it until Copernicus. The Indian tradition reached the same structural conclusion independently. The Surya Siddhānta (the authoritative astronomical text that preceded and ran alongside Āryabhaṭa's work) calculated the Sun's diameter at 6,500 yojanas against the Earth's 1,600, placing the Sun at roughly four times the Earth's diameter. The numbers differed from Aristarchus's estimate. The conclusion did not: the Sun was physically larger. Both traditions absorbed that conclusion without rupture, because the computational system worked regardless of which body was physically larger. Physical evidence pointing toward heliocentrism was present in both traditions long before Nīlakaṇṭha. It was not sufficient. A system that computes correctly without resolving the physical question has no internal reason to resolve it.

What Any Path to Heliocentrism Required

The move from geo-heliocentrism to full heliocentrism – placing the Sun at the physical center and accepting that the Earth orbits it – could not be compelled by any observation available to naked-eye astronomy. All ancient astronomical observation is necessarily geocentric in its physical frame: you stand on Earth, you measure angular positions relative to Earth, you build models that reproduce what you see from Earth. The question of whether Earth itself moves in orbit around the Sun cannot be settled from that standpoint. Detecting Earth's orbital motion requires measuring stellar parallax, the apparent shift in a star's position as Earth moves from one side of its orbit to the other. That detection requires instruments of a precision unavailable until Friedrich Bessel in 1838. Kepler had no direct observational evidence that Earth moves. He had a derived conclusion: the orbit of Mars refused to fit any circle, which forced the Sun to be the physical center of planetary orbits, from which Earth's orbital motion followed as an implication. That derived conclusion required the Mars residual. The Mars residual required Tycho's instruments.

This matters for understanding Nīlakaṇṭha's position specifically. The Aryabhatan tradition had already accepted something radical: that the Earth rotates on its axis. Āryabhaṭa stated it. Nīlakaṇṭha's tradition carried it. On paper, this should have been a head start. A tradition in which Earth's motion had been a thinkable claim for a thousand years, surviving Brahmagupta's sustained rebuttal without being suppressed, had already crossed the cultural threshold that, in seventeenth-century Europe, would put Galileo before the Inquisition. Half the heliocentric commitment was, in this sense, in hand. But Earth's rotation and Earth's orbital motion around the Sun are different claims requiring entirely different evidence. Rotation, the Earth spinning on its axis, explains the daily motion of the stars and is compatible with geocentrism, geo-heliocentrism, and full heliocentrism alike. Orbital motion, the Earth revolving around the Sun, requires demonstrating that the Sun is the physical center of planetary paths, a demonstration that appears only at arcminute precision in the behavior of planetary orbits. The tradition's openness to Earth's rotation gave it no path to Earth's orbital motion. These are distinct questions answered by distinct data.

There is a deeper question worth naming. Āryabhaṭa had already accepted Earth motion of one kind — rotation — despite no direct observational evidence, on the strength of a conceptual argument about relativity. Why did Nīlakaṇṭha, having reorganized the planets around the Sun, not extend Āryabhaṭa's logic and propose that the Earth moved too?

Three reasons work in concert. Āryabhaṭa's rotation explained an observed phenomenon, the westward motion of the stars, by relocating motion into the observer. Nīlakaṇṭha's reorganization solved a technical problem, the latitude asymmetry for Mercury and Venus. Neither required Earth's orbital motion to be solved; the leap would have answered no open question. Beyond that, rotation and orbital motion are conceptually asymmetric in a way easy to miss. Rotation has an everyday analog — the boat-and-river image transfers naturally. Orbital motion has none. Worse, orbital motion predicts stellar parallax, an apparent shift in stellar positions that would not be observed until 1838. To accept orbital motion, one had also to accept that the stars are at distances so vast that parallax becomes too small to measure: a layered commitment Āryabhaṭa's rotation did not require.

The deepest reason is the tradition's orientation. Indian astronomy was problem-solving, not cosmologically speculative. Greek natural philosophy produced thinkers — Philolaus, Heraclides, Aristarchus — who proposed Earth motion as a philosophical move untethered from immediate utility. The Indian tradition did not. Speculation about the underlying physical arrangement of the cosmos, separated from the working calculation, was not what the tradition rewarded. Āryabhaṭa was already an outlier for proposing rotation, and the tradition could not agree about him for the next thousand years. Nīlakaṇṭha was not going to make the additional speculative leap, unsupported by his problem, by any analog, or by the missing parallax.

The available observational data, moreover, did not make the asymmetry intolerable. This is not because the Kerala tradition was observationally passive. Parameśvara made systematic eclipse and planetary observations for fifty-five years, comparing computed positions against observed ones and revising parameters where they diverged. The tradition took observation seriously because it had to: jyotiṣa, the discipline these astronomers served, governed the most consequential decisions in people's lives. Births, marriages, coronations, the timing of battles and religious ceremonies were all determined by planetary positions. Getting those positions wrong had real consequences. The precision the tradition demanded was therefore real and substantial.

An eclipse prediction that placed the shadow at the wrong moment, or at the wrong degree of totality, threw the accompanying ritual into disorder. The priest could not perform the correct observances at the correct instant because the instant had been wrongly calculated. The pressure on the tradition was therefore real and intense, but it was pressure for computational accuracy, not for physical truth. The court did not ask why the eclipse happened. It asked when. And as long as the when was right, the question of why was never forced open.

The instruments available were calibrated to exactly that precision: accurate to perhaps five to ten arcminutes, enough to determine planetary conjunctions, eclipse timings, and the nakṣatra of the moment. This is not because the tradition's thinkers lacked ambition for greater precision. It is because the instruments had not yet been built to that scale. At that precision level, discrepancies appear and are real, but they are small enough that refining the parameters makes them disappear. The framework survives each encounter with the data because no single residual is large enough and stubborn enough to resist correction.

The scale of instrumentation that would have made such a residual undeniable (arcminute-level precision, achieved only when a specific competitive agenda created the incentive to fund it) was not present in Kerala. That is the subject of the next chapter.

Vasco da Gama landed at Calicut in 1498, three years before Nīlakaṇṭha completed the *Tantrasaṅgraha*. The lineage that had run unbroken from Mādhava to Nīlakaṇṭha does not continue past the sixteenth century. What the disruption foreclosed was not the next breakthrough: the conditions for that breakthrough did not yet exist in any tradition, and this chapter has argued why. What it foreclosed was the trajectory along which those conditions might have been built. Parameśvara had sustained a fifty-five-year observational program; Mādhava had taken computational precision well past anything jyotiṣa required. The discipline existed. Whether, in another century, that discipline would have been turned toward instruments at arcminute precision — the only path that could have produced a Kerala Mars residual — is unanswerable. That unanswerability is the historical loss. Not a breakthrough denied, but a trajectory ended before it could be tested.

The first thing Kepler had that Nīlakaṇṭha did not was observational data of a different order of precision. Tycho Brahe's program at Uraniborg produced planetary observations accurate to within one or two arcminutes. This was not merely a quantitative improvement: it was a change in what the data could show at all. Errors that had been invisible or absorbed as rounding became, at that precision, impossible to dismiss.

The second thing Kepler had was an institutional context that demanded physical explanation rather than computational adequacy. The stakes were not merely computational: they were cosmological. What is the actual structure of the solar system? That question was live, contested, and considered urgent. In that context, a small discrepancy between theory and observation was not merely a correction to be made. It was a potential verdict on the entire framework.

The step from geo-heliocentrism to heliocentrism did not require new mathematics. It required a situation in which the existing mathematics could no longer protect the model from the data. What was missing was an intellectual environment in which residual error became a verdict rather than an acceptable approximation. The conditions made rupture necessary. Kepler made it happen. Without those conditions, another Kepler would have computed orbits and moved on. Without Kepler, those conditions would eventually have produced someone else. Structure and individual are not alternatives. The structure sets the stage. The individual walks onto it.

The Name and the History

All of this explains why the breakthrough came from Europe, not from Kerala. It does not yet say where Nīlakaṇṭha belongs in the story when the absence of that breakthrough has been accounted for.

Consider what his achievement required of him. The comparable position in European astronomy (the Tychonic geo-heliocentric model) was a tandem achievement. Tycho Brahe spent twenty years at a royal island observatory, with state funding and purpose-built instruments accurate to within one arcminute, assembling the most precise planetary database in history. Johannes Kepler then spent years analyzing that inherited data, extracting from it the forcing anomaly that made the geo-heliocentric model untenable. Two exceptional minds, a royal patron, a dedicated institution, decades of precision measurement, and the pressure of a live cosmological competition: all of this converged to produce the Tychonic position. Nīlakaṇṭha had none of it.

He had a lineage tradition, instruments calibrated for practical purposes, and the mathematical inheritance of Mādhava and Parameśvara. He arrived at the same structural position by following what the śīghra corrections implied: by asking, with mathematical precision and intellectual courage, what the computation was actually saying about the physical structure of the solar system. Aristarchus had reasoned from physical size; Copernicus would reason from cosmological aesthetics; Nīlakaṇṭha reasoned from inside the mathematics. He let the computation make the cosmological claim. That is not a specialist's footnote. It is one of the most extraordinary feats of mathematical inference in the history of astronomy.

I grew up knowing Āryabhaṭa's name. I did not know Nīlakaṇṭha's, not until I started following this thread. That asymmetry is itself part of the story. The achievement, largely unknown outside specialists in Indian mathematics, does not carry the name the way the breakthrough does. Copernicus traveled; Nīlakaṇṭha remained a specialist's reference. The book has argued why, and I believe the argument. But arriving at this moment in the account, I want to note what was there: a mind working at the edge of what was knowable, in a tradition that had given it extraordinary tools, arriving at something true and structurally significant, and a tradition that received it as one more successful step and moved on.

There is a further question. Did Nīlakaṇṭha know what he had done? Technically, yes — his commentaries show clear awareness of the latitude solution and the structural reorganization. But did he understand himself as the person who had established that all five planets orbit the Sun? Almost certainly not. He was inside the framework that absorbed his own result, and from within that framework there was no signal that something cosmologically decisive had occurred. He solved specific problems. He solved them brilliantly. He had no way to know, from within the tradition's self-understanding, that he had also done something else entirely.

The texts were written in Sanskrit, the pan-Indian scholarly language, in principle readable by any educated scholar from Kerala to Kashmir. But Sanskrit guaranteed readability, not circulation. Texts traveled through two mechanisms: manuscript copying carried by traveling scholars, and adoption into the curriculum of major institutional centers. The great northern centers at Varanasi, Ujjain, and Navadvīpa had their own working traditions, their own lineages, their own computational systems. They had no competitive reason to seek out Kerala's results, and Kerala's lineage-based tradition made no particular effort to send them. Demand drives circulation. The demand did not exist.

There is a quiet test of this claim, three and a half centuries later. In 1869, an astronomer named Candraśekhara Sāmanta (working in Odisha, trained entirely within the traditional Indian astronomical lineage) wrote a treatise he also titled *Siddhāntadarpaṇa*, in which he proposed a model in which the five planets orbit the Sun. He arrived at the geo-heliocentric architecture independently, working from purely traditional sources, in a different region of India, in a different lineage, in the second half of the nineteenth century. The internal conditions that had produced the geo-heliocentric model in Kerala in 1501 were durable enough to produce it again, in Odisha, in 1869. The external conditions that would have circulated it had still not arrived. Sāmanta's result, like Nīlakaṇṭha's, is known today only to specialists in the history of Indian astronomy. The same internal architecture, twice. The same silence around it, twice.

The standard narrative of planetary astronomy compresses centuries into a single line: Greek astronomy, Islamic refinement, Copernicus, Kepler, Newton. That compression obscures the existence of an independently developed geo-heliocentric tradition in Kerala that, decades before Tycho, had produced a structurally equivalent planetary model through mathematical inference alone — and the Odisha echo three centuries later that reached the same architecture independently again. The name belongs in the history of astronomy not as a footnote to a story about what did not happen, but as a figure who did what was achievable with what was available, and whose achievement, measured against the conditions that produced it, stands among the most extraordinary in the record. The history of planetary astronomy has not fully absorbed what that means. It should.

Chapter Ten

When Error Becomes Evidence

Kepler, the Forcing Anomaly, and the Answer the Tradition Could Not Reach

The remaining transformation, from geo-heliocentrism to full heliocentrism, was completed not in Kerala, but in early 17th-century Europe, where a German mathematician inherited the most precise planetary data ever assembled and refused to let the discrepancy go.

The Kerala School's mathematics in some specific domains was comparable to — and occasionally exceeded — what European astronomers of the same period possessed. What broke the impasse was not a matter of superior talent or intellectual openness: Galileo's fate suggests that parts of European intellectual life were hardly a haven for heterodox ideas. What broke was more specific: a single planetary orbit refused to behave. An error small enough to ignore decided the model. And the conditions surrounding that error (the precision of the data, the urgency of the theoretical competition, the institutional demand for physical explanation rather than computational adequacy) made it impossible to absorb.

What this chapter examines is the specific structure of a forcing anomaly: what it is, what produces it, and why a working system rarely generates it from within itself.

The Data That Changed Everything

The story begins not with Kepler but with Tycho Brahe, a Danish nobleman who spent two decades at Uraniborg, his observatory on the island of Hven, doing something no one had done before at sufficient scale: measuring planetary positions with systematic, sustained, arcminute-level precision.

The full Moon, as seen from Earth, spans about half a degree, or roughly thirty arcminutes. Tycho's instruments could locate a planet to within one or two of those tiny divisions. No previous sustained observational program had come close to this.

Previous astronomical observations were accurate enough for most computational purposes. Parameśvara's fifty-five years of eclipse records in Kerala (made with a gnomon and armillary sphere) could show that the Parahita parameters were drifting. That was precision enough to demand a response. It was not precision enough to expose a residual the framework could not absorb. At Tycho's level of precision, discrepancies that had been invisible became undeniable.

Tycho's program at Uraniborg was the product of specific conditions. He had royal patronage directed at precision measurement as an end in itself, not at computation for calendrical purposes. He had an institutional culture that treated the gap between theory and observation not as a nuisance but as information. And he had a theoretical motivation: the competition between Ptolemy's geocentric system, Copernicus's heliocentric system, and his own hybrid model had made planetary data not merely useful but urgently contested. Better observations could, in principle, settle a live dispute.

The instruments that made this possible were themselves the product of a specific developmental trajectory. Tycho's mural quadrant at Uraniborg was built into solid masonry walls, two meters in radius, sheltered from wind in underground observing rooms. His armillary spheres extended to nearly three meters. These were not refinements of existing instruments: they were a qualitative leap in physical scale that had taken European astronomical instrumentation a century of incremental building to reach. Scale is what determines angular precision: a two-meter quadrant can be subdivided to arcminute accuracy; a portable wooden armillary sphere cannot. Tycho's program produced data accurate to within one or two arcminutes not because he was more careful than his predecessors but because the physical infrastructure he built made that precision achievable for the first time in history. The tool created the crisis. Without it, the eight-arcminute residual in Mars's orbit would have been absorbed into measurement uncertainty.

What a Forcing Anomaly Is

A forcing anomaly is an error that is visible, persistent, and treated as a verdict, one that cannot be absorbed without confronting the possibility that the model itself is wrong. These three conditions must hold simultaneously.

Visibility requires data precise enough to surface the error. Persistence means the error cannot be made to disappear by adjusting parameters within the existing framework. Treatment as a verdict rather than a nuisance requires an institutional context that demands physical explanation rather than computational adequacy.

Remove any one of the three conditions and the anomaly is absorbed. Imprecise data makes the error invisible. A flexible framework absorbs it through refinement. And a prediction-oriented tradition treats it as an acceptable approximation.

There is a fourth failure mode, which underlies the others: the data can be precise but not trusted. A residual that can be attributed to instrument error, atmospheric uncertainty, or observational slippage is absorbed before it can become a verdict. For the anomaly to be undismissable, the precision regime that produced it must be institutionally established: its margins known, its discipline recognized, its results impossible to attribute to artifact without confronting the measurement itself.

The Decision About Mars

When Tycho Brahe died in 1601, he left his data to Kepler (then his assistant) along with an unfinished problem: Mars. The orbit of Mars was the most stubborn anomaly in the Tychonic data. Every model produced a residual. Every parameter adjustment that improved the fit at one point made it worse at another. Mars refused to behave.

Tycho himself, had he lived, would likely have remained at geo-heliocentrism. What ended it was not the data alone but Kepler's specific refusal to absorb the Mars residual. Copernicus proposed. Kepler forced. That distinction is the whole of the difference.

Mars was not simply the hardest case. It was the specific planet whose orbit cannot be reconciled with any circular model at arcminute precision. Mercury and Venus are naturally solar: their organization around the Sun had been stable for centuries, including in Nīlakaṇṭha's model. Mars is different: its orbital eccentricity is large enough that circular models fail exactly where precise data reveals the failure. Mars is where the geo-heliocentric compromise breaks. Without Mars, the Tychonic architecture survives.

The reason is specific. Every planetary orbit is in reality an ellipse. Every planet produces some departure from the circular motion that Ptolemy's machinery assumes. But Ptolemy's machinery is not naive. Its equant — an off-center point around which the planet moves at uniform angular speed — already accounts for the apparent variation in a planet's speed across its orbit, which is the first thing an elliptical orbit produces. What the equant cannot capture is the orbit's actual spatial shape: how squashed the ellipse is compared to a circle. That residual deformation does not scale with eccentricity alone. It scales with eccentricity squared. For nearly circular orbits, the squared term is tiny and the deformation vanishes. Only at higher eccentricities does it become large enough to see. Mars's eccentricity of 0.0934 produced a post-equant residual roughly four times larger than Jupiter's or Saturn's, and more than two hundred times larger than Venus's essentially circular orbit. At opposition, when Mars is closest to Earth and Tycho's measurements most precise, that residual reached eight arcminutes, four times the noise floor of his instruments. Jupiter and Saturn produced residuals of two to three arcminutes, right at the detection limit: real, but deniable. Venus's residual was buried entirely. Mercury's eccentricity is larger than

Mars's, but Mercury is observed only at twilight, when observations are too brief and noisy for any conclusion. Mars was the only planet whose orbit was eccentric enough to produce an undeniable residual, visible at close range under Tycho's best observing conditions. Every other planet stayed quiet. Mars could not.

Kepler's trust in this data was not blind. He had worked directly under Tycho at Benatky, near Prague, where Tycho had relocated with his data and instruments after leaving Uraniborg. He knew the instruments, knew their margins, knew what they were worth. When he said the eight-arcminute residual was real and not instrument error, he was not trusting a stranger's numbers. He was applying his own first-hand knowledge of what those instruments could and could not do. His faith in the data was earned faith.

That earned trust is itself a structural condition, not a personal one. Many astronomers had access to Tycho's data after his death. What made the Mars residual undismissable for Kepler specifically was that he knew too much about the instruments to treat the eight arcminutes as noise. He had calibrated those margins himself. He could not pretend the discrepancy was there by accident. The precision regime had made absorption intellectually impossible for anyone who understood it.

Kepler made a decision that sounds almost unreasonable when stated plainly: he narrowed his entire effort to a single planet and refused to let an eight-arcminute discrepancy go. Eight arcminutes (less than a sixth of a degree) was small enough that earlier astronomers would have ignored it. Kepler did not.

His reasoning was precise. Tycho's observations were accurate to within one or two arcminutes. An eight-arcminute residual was therefore not measurement error. It was signal.

And once treated as signal, it changed the status of the model entirely. The problem was no longer to improve the fit, but to explain why it failed. The model was not slightly off. It was wrong in a way the data made visible.

As someone who has run mutation screens in a genetics laboratory, I recognize this reasoning precisely. In experimental science, the forcing result is the one you cannot explain away. You have adjusted every parameter, checked every control, run the experiment again, and the number still does not fit. At that point you face a choice: absorb the discrepancy and move on, or treat it as a verdict. A tradition oriented toward computational adequacy absorbs it. A tradition demanding physical explanation cannot.

The competition between Ptolemy, Copernicus, and Tycho made the stakes cosmological, not merely computational. The question was not how to produce accurate predictions: all three systems could do that reasonably well. The question was what the solar system actually was. In that context, an eight-arcminute residual was not a nuisance. It was a verdict.

A Different Kind of Answer

What followed Kepler's decision to treat the Mars residual as signal rather than noise was years of calculation: patient, laborious, sometimes despairing work through the Tychonic data. He tried every variation of circular motion. He tried eccentric circles, equants, combinations of parameters. Nothing worked. Mars continued to misbehave.

The answer, when it came, was not a refinement of the existing framework. It was a replacement of its central assumption. The eight arcminutes had already made that clear: no adjustment within circular motion could remove the error.

Planetary orbits are not circles. They are ellipses, with the Sun at one focus.

It is easy to hear this as merely a geometric substitution: replace one shape with another, adjust the parameters, done. What actually happened was more consequential. The circle had been the fundamental shape of celestial motion since Ptolemy. It was not a parameter. It was the assumption Kepler questioned, and replaced because the data gave him no choice. Not reduced. Eliminated.

More importantly, the ellipse was not merely a better approximation. It was a different kind of answer. A circle with adjusted parameters is still a circle. An ellipse is a different framework. It restructured the entire planetary system around a physical reality: the Sun sits at one focus of every planetary ellipse, not at the geometric center. The Sun is not just computationally central, as it had been in Āryabhaṭa's corrections and Nīlakaṇṭha's model. It is physically central, the anchor of actual motion.

That is what the Indian tradition's framework could not produce from within itself. Not the ellipse: the mathematics for ellipses was available. Not the geo-heliocentric model: Nīlakaṇṭha had independently arrived at that same geo-heliocentric architecture decades before Tycho reached it in Europe. What the tradition could not produce was the forcing anomaly that made the geo-heliocentric model itself untenable: the specific, stubborn, undeniable residual that would force the Earth off its center.

What the Comparison Reveals

The comparison between Nīlakaṇṭha and Kepler is not a comparison of talent or insight. Both were working with the best tools available to them. Both recognized the Sun's structural importance to planetary motion. Both were capable of the mathematics their respective advances required. Kepler did not have better mathematics than Nīlakaṇṭha. He had a worse tolerance for error, and a tradition that made that intolerance productive rather than merely frustrating.

Individual insight appears in both traditions. What differs is whether the surrounding structure forces those insights to connect. In Kepler's case the theoretical landscape was so actively contested that a small anomaly could not be set aside without choosing between irreconcilable frameworks. That is the condition the Indian tradition never encountered in the same form.

A careful reader might object here: did not Brahmagupta himself demand physical explanation? He did, and the tradition had genuine competitive debate. Brahmagupta attacked Āryabhaṭa sharply and publicly, and the rivalry between the *Ārya-pakṣa* and *Brāhma-pakṣa* schools, the Aryabhatan and Brahmaguptan camps, continued for centuries. Competition was present.

But there is a structural difference between competition within a shared framework and competition between irreconcilable frameworks. Brahmagupta was competing to settle a debate about one claim inside a common computational system. Copernicus, Tycho, and Kepler were competing across frameworks that could not coexist, each trying to show that the other's entire cosmological architecture fails. In that environment, finding where the rival model breaks was the primary intellectual currency. A discrepancy was not a local error to fix. It was a weapon. The tradition produced two near-misses of different kinds: Brahmagupta naming a real physical force and using it to close a debate, Nīlakaṇṭha reorganizing the entire planetary system around the Sun and having it received as a refinement. Both insights were genuine. In both cases the tradition found them useful before they could become generative.

Competition inside the same framework sharpens: it produces better parameters and deeper precision within the existing structure. Call this absorptive pressure: the system absorbs the challenge and emerges more refined. Competition between incompatible frameworks breaks: it forces the question of which structure is right. Call this forcing pressure: the system cannot absorb the challenge without confronting the possibility that the structure itself is wrong. Forcing pressure tends to emerge where three conditions converge: irreconcilable frameworks are in active competition; precision tools are sharp enough to expose stubborn residuals; and the stakes — theological, institutional, political — are high enough that the unresolved anomaly cannot be ignored. These conditions are not guaranteed to produce rupture. But without them, rupture rarely occurs.

Scientific revolutions do not occur simply because better ideas are possible; they require conditions that make existing ideas untenable: error must become both visible and unacceptable, and the surrounding framework must make acceptance of that verdict possible. Nīlakaṇṭha's model was not untenable. It worked. It produced better predictions than its predecessors. There was nothing to force it beyond itself.

♦

In the early 18th century, more than two centuries after Nīlakaṇṭha, the Rajput ruler Jai Singh II built a series of large-scale observatories across India: at Delhi, Jaipur, Ujjain, Varanasi, and Mathura. The Jantar Mantar observatories were instruments of precision measurement on a scale the earlier tradition had never had: massive masonry instruments designed to determine planetary positions with arcminute accuracy.

Jai Singh's program was an attempt to supply what had been missing — the precision observational infrastructure that might have surfaced the kind of discrepancies Tycho found. He was also aware of European astronomy and corresponded with Jesuit astronomers. The program came too late to influence the development of the tradition in the way that Tycho's work influenced Kepler. But it is a remarkable acknowledgment, within the tradition itself, that scale of observation mattered, that better data was not merely useful but necessary.

What Jai Singh could not supply, by the 18th century, was the theoretical landscape that would have made the data forcing rather than merely more precise. The competition between cosmological frameworks that made Kepler's anomaly urgent had been settled. The ellipse was known.

Nīlakaṇṭha had the model. Jai Singh had the instruments. Neither had the combination of conditions — precision data, theoretical urgency, institutional demand for physical explanation — that such a verdict requires.

Kepler made the model untenable. Newton made it intelligible. That distinction matters. Kepler showed that planetary orbits are ellipses and gave the rules that governed them, but he could not explain why the rules held, only that they did. Newton supplied the causal mechanism: gravity, operating at a distance, producing exactly the elliptical orbits Kepler had described. That residual broke one framework; the gravitational synthesis replaced it with one that explained not just the data but the reason for the data. Each step required the previous one. Nīlakaṇṭha had built the full geo-heliocentric model — through mathematical inference rather than Tycho's precision measurement program — and had done so before Tycho's time. What came next required not just the model but the anomaly that made the model insufficient, and that required conditions neither Kerala nor Jai Singh possessed.

♦

Indian astronomy absorbed the anomaly. Kepler let it become a verdict.

The thread that began with a verse about triangular numbers has now run its course. From Bhāskara II's pedagogy to Āryabhaṭa's computation to the rotation claim to Nīlakaṇṭha's geo-heliocentric achievement to the conditions that made Kepler's question possible: what the thread reveals is not a hierarchy of traditions but a map of the conditions under which knowledge advances, stalls, and sometimes breaks through.

Chapter Eleven

What the Map Shows

A Set of Conditions, Not a Single Mechanism

The verse that started this book is still where it started. What has changed is how to see it.

A twelfth-century Sanskrit verse encoding two rules of summation in a form that can be memorized and transmitted without writing: this is not ornamentation. It is infrastructure. The tradition had solved a specific problem: how do you keep a precise mathematical rule intact across centuries without print? You encode it in a language whose grammar enforces precision and make it beautiful enough to carry in memory. The verse is simultaneously a poem and an algorithm.

I began this book by looking for mechanism. What I found, following that thread through ten chapters, was not a single mechanism but a set of conditions: conditions that shaped what the tradition could achieve, what questions it was oriented toward, and where it reached and where it stopped. That set of conditions is the map this book has been building. What follows is the map in its simplest form.

The map resolves into three features.

The first is a methodological principle that applies to any claim about intellectual history, in any direction.

Mathematical equivalence is not historical evidence. To establish transmission you need the chain: named intermediaries, datable texts, traceable transformations. The standard must hold in all directions. This is not skepticism about achievement. It is respect for it.

The second is a set of structural conditions that shaped what the Indian astronomical tradition could and could not do, and that generalize beyond it.

Authority, distributed rather than concentrated, changes what a thinker can compose and transmit without fear. Āryabhaṭa set down at twenty-three that eclipses are shadows and the earth rotates, in a world of sacred cosmology, without institutional punishment. The tradition surrounding him argued fiercely: Brahmagupta spent pages refuting him, but argument is not silencing. A tradition that can argue without silencing preserves the possibility of being wrong in public, which is the precondition for being corrected rather than merely condemned.

This is not a claim of uniform openness on one side or uniform closure on the other. The Vedic world had its own domains of enforced conformity: in social structure, ritual practice, and linguistic purity, distributed authority could police with considerable force. And Europe had its own protected intellectual spaces: universities, princely courts, the Republic of Letters, environments where heterodox thinking found shelter when it served the right interests. The structural argument is not about which civilization was more tolerant in general. It is about which specific institutional arrangements, in which specific domains, created the conditions for a particular kind of knowledge to be stated, tested, and extended without fear.

The same structural condition that gave Āryabhaṭa the permission to be right — the absence of a single authority with both the power and the incentive to enforce one answer — is structurally connected to the absence of the forcing pressure that Kepler's tradition possessed. Distributed authority prevents censorship. It also means there is no institution able to force irreconcilable frameworks into collision: the kind of collision that, in Europe, made individual brilliance produce forcing anomalies. The permission to be uncertain, which allowed Āryabhaṭa to leave physical tensions unresolved in his text and move on, is the same permission that meant Nīlakaṇṭha's geo-heliocentric achievement could be received as a success rather than as an incomplete step demanding continuation.

The freedom and the limit are not opposites.

They are produced by the same conditions.

The reward structure that came with distributed authority reinforced this. Reputation was built by making the system work better: improving constants, sharpening methods, extending results within the established framework. Brahmagupta improved Āryabhaṭa's parameters and became one of the tradition's towering figures. No comparable reward existed for showing that the framework itself was wrong. The competition that existed — and it was real, sustained, and fierce — was competition within a shared framework, not between irreconcilable ones. That distinction is the difference between refinement and rupture.

Institutional orientation shapes not just what answers get found but what questions get asked. A tradition organized around computational accuracy for calendrical and ritual purposes will produce extraordinary mathematical tools. But institutional orientation is not a ceiling on individual brilliance: the Kerala tradition consistently produced minds that exceeded what the institution required. Mādhava's mathematics, Parameśvara's fifty-five years of observation, Nīlakaṇṭha's restructuring: none of these were institutionally demanded. What the institution shapes is the competitive landscape: whether individual brilliance collides across irreconcilable frameworks or develops within a shared one.

The third is a way of seeing that applies beyond this book's particular subject.

When you encounter a claim about what a tradition achieved — or failed to achieve — four questions will carry you further than admiration or dismissal.

What were the conditions for public expression? Not just whether dissent was formally permitted, but whether the institutional structure created the distributed authority that allowed a thinker to be publicly wrong without being silenced. Permission is not a binary. It has a structure, and the structure has consequences.

Āryabhaṭa is the positive case: argument without silencing, because no single authority held both the power and the incentive to convert disagreement into punishment. Galileo is the negative one. The difference is not the claim. It is the structure the claim encountered.

What counts as evidence for transmission? Is there a chain: named intermediaries, datable texts, traceable transformations? Or is the argument resting on equivalence alone? Equivalence is real. It reveals the unity of the underlying mathematics, or physics, or logic. What it cannot reveal, by itself, is the path of ideas.

The Hindu-Arabic numerals are the clean case: named intermediaries, datable texts, explicit acknowledgment at every link. The Kerala calculus is the incomplete case: sender known, channel plausible, receipt missing. Both cases are real. Only one meets the standard.

What questions was the tradition oriented toward? The orientation shapes the questions, and the questions shape what kinds of anomalies become visible and what kinds remain tolerable. A tradition oriented toward computational adequacy will handle discrepancies differently from one demanding unified causal explanation. Neither orientation is superior in the abstract. Each produces different achievements and different limits.

Āryabhaṭa asked how to make the numbers work. Kepler asked why they work that way. Those are not two answers to the same question. They are different questions, and a tradition finds what it is looking for, and does not find what it is not looking for.

What would have forced rupture? A working system does not break from within. Breakthroughs require something to break the existing framework: an error the framework cannot absorb, arising from precision data, theoretical urgency, and the competitive pressure that makes a small discrepancy into a verdict rather than a correction. If those conditions are absent, the framework continues to work, and the tradition continues to develop within it, and the question of whether a breakthrough was "possible" is less illuminating than the question of what conditions its possibility required.

Nīlakaṇṭha had the mathematics. What he did not have was Kepler's eight arcminutes: the specific, stubborn, visible error that could not be absorbed. Without that, the model continued to work.

The geo-heliocentric model was not an incomplete solution. It was complete — complete enough that the Sun's computational centrality and the Earth's experiential centrality could coexist without contradiction. Tycho arrived at the same equilibrium through a different path and remained there. What Kepler possessed that neither had was a residual the data refused to absorb, one that made the computational and the physical impossible to keep separate.

♦

The tradition produced Āryabhaṭa's computation and Bhāskara I's sine approximation and Mādhava's infinite series and Nīlakaṇṭha's geo-heliocentric model. These are genuine achievements. They stand without inflation. What the framework adds is not credit or blame but precision: about what the achievements were, what conditions made them possible, what conditions were absent when the tradition stabilized, and why the limit was structural rather than personal.

The tradition's computational power was not a claim made by its practitioners. It was measured, by outsiders, against the best available alternatives.

The French astronomer Guillaume Le Gentil, who reported on Indian astronomy from Pondicherry between 1768 and 1770, compared two predictions of the August 1765 lunar eclipse against the eclipse itself. A Tamil astronomer using methods derived from Āryabhaṭa's tradition was off by 41 seconds. The best European astronomical charts of the era (Tobias Mayer's tables, the standard reference of the time) were off by 68 seconds. Le Gentil had no stake in the outcome. He reported what he found.

Sixty years later, in 1825, Lieutenant-Colonel John Warren of the British East India Company encountered a calendar maker in Pondicherry who calculated a lunar eclipse using no instruments, no books, and no written tables. He arranged shells on the ground and worked from mathematical tables memorized in verse, exactly the transmission mechanism this book described in its first chapter. His prediction was off by no more than four minutes. The tradition that Āryabhaṭa had compressed into 121 verses in 499 CE was still being carried in human memory thirteen centuries later, still producing results, still working.

This is what the conditions this book has described actually produced. Not a monument. Not a claim. A calendar maker in Pondicherry in 1825, computing an eclipse with shells on the ground, from verses in his head, within four minutes of the truth.

It is the system at its absolute peak, and at the same time its boundary.

The computation is exact enough to predict the shadow and grounded enough to explain it. The eclipse is not a mystery in this system; it is geometry — alignment, shadow, position — fully intelligible within the framework. But the geometry did not displace what came before.

The most vivid illustration of this is not in the mathematics. It is in the astrology that continued alongside it. Rahu and Ketu (the celestial demons responsible for eclipses in the pre-Aryabhatan cosmology) did not disappear when geometry replaced them. They were reinterpreted. In the post-Aryabhatan tradition, Rahu and Ketu became the lunar nodes: the precise mathematical points where the moon's orbit intersects the ecliptic, at which eclipses become geometrically possible. The demon became a coordinate. The coordinate continued to function as a demon. Today, every Indian astrological chart includes Rahu and Ketu, their positions calculated with the same precision as any other planetary body, their influence consulted for auspicious timing, their transits watched during eclipse seasons. The priest explaining Rahu's malefic influence is not contradicting Āryabhaṭa. He is operating downstream of Āryabhaṭa, using a term that now carries both geometric precision and ritual weight, with no awareness that there was ever a conflict to resolve. Āryabhaṭa did not defeat Rahu. The tradition gave Rahu a new address and kept him in service. This is absorptive pressure operating at its most sophisticated: a framework that can redefine its own symbolic furniture without losing a single piece of ritual furniture. The geometry was accepted. The demon survived. And the framework that could accomplish that had, by the same token, no internal pressure to extend the geometry further than the computation required.

The same absorptive capacity operated when Nīlakaṇṭha organized all five planets around the Sun. The Sun became computationally central; the Earth remained cosmologically fixed; the tradition absorbed both without resolving the tension between them. The Rahu case and the geo-heliocentric case are the same mechanism at different scales of the framework.

A tradition that can account for the eclipse through geometry does not, in this form, require that all planetary motion be governed by a single, unified physical model. The numbers work, the local explanations hold, and nothing in the structure demands that they be brought under one consistent account.

That distinction — between a system that explains where needed and one that must explain everything — is the line this book has been tracing. The Indian astronomical tradition crossed the first with extraordinary success. It did not, for structural reasons, cross the second.

Not for lack of intelligence or insight, but because the conditions that turn a working system into an insufficient one (precision that exposes error, institutions that demand explanation, anomalies that cannot be absorbed) did not converge.

The calendar maker is therefore not a contradiction of the argument but its final demonstration. He represents the full maturity of a tradition that solved, with elegance and durability, the problem it set itself: how to compute the heavens with precision and continuity across centuries. He also represents the absence of the specific pressure that would have forced the next question, not because the question could not be asked, but because nothing in the system required it to be.

Scrutiny did not diminish the tradition. It revealed what admiration alone could not: a set of minds working with exceptional care on genuinely hard problems, within conditions that shaped both what they could reach and what they could not, producing results that are more interesting under examination than under praise.

The verse is still where it started. The frame around it is what has changed. And the frame is portable.

A tradition does not fail for lack of intelligence, but for lack of conditions that make intelligence dangerous: the permission to be wrong in public, the pressure that makes a sufficient system feel like it has more to answer for, the competitive pressure that refuses to let error go unexamined. Those conditions can be built. They can also fail to be built. In every case examined in this book — Āryabhaṭa, Brahmagupta, Nīlakaṇṭha, Kepler — the intelligence was present. What differed was the pressure.

♦

One responsibility a framework has to its readers is to name what it does not claim.

This framework describes one pathway by which knowledge systems change: an anomaly that becomes visible, persistent, and impossible to dismiss, within a structure that cannot absorb it. It does not describe all the ways institutions change. External shocks reorganize institutions through pathways that have little to do with anomaly absorption: wars reshape universities, technological displacement obsoletes whole fields, political collapse rewrites the institutions that did the absorbing. Nor does it describe all routes to new understanding. Cantor's transfinite set theory expanded mathematics without breaking an established framework; there was no equivalent of the Mars residual. Some breakthroughs are syntheses rather than ruptures. The framework explains the rupture case, which is the cleanest case in scientific revolution. It does not claim that all advance follows this path.

The three conditions are necessary, not sufficient. A finding can meet all three and still not produce timely change. The replication crisis in psychology has been visible, persistent, and treated as a verdict by many practitioners for more than a decade; methodological reform has come slowly. Climate science is sharper: the scientific anomaly has been settled for decades, and the political and institutional response is still being deferred. The framework explains why working systems absorb anomalies. It does not guarantee that, once the conditions are met, the response will be commensurate. Political, economic, and cultural forces can override the structural logic. The conditions describe what makes change possible, not what makes it inevitable.

The framework is descriptive, not prescriptive. It explains when systems are likely to revise themselves. It does not claim that all working systems should. Some stable systems are correctly stable: Kerala's geo-heliocentric model was the right answer under the conditions available, and naming the structural reasons for that stability is not the same as judging stability as failure.

The case base is small. The argument is built on a few well-documented cases: Indian mathematical astronomy and the European astronomical revolution. The generalization to other domains is analogical and has not been systematically tested. Chinese astronomy under the bureau system, Islamic astronomy after Ulugh Beg, the slow acceptance of plate tectonics, the history of legal and religious institutions: each is a plausible test case. Some would likely support the framework. Some may not. The book argues from where the evidence is densest. It does not claim those patterns generalize without further testing.

The argument also rests on what the historical record shows, and what the record shows is itself a function of what survived. Traditions that suppressed dissent also tended to suppress evidence of suppression. The absence of documented punishment is consistent with the structural argument, but it cannot rule out the possibility that the record is incomplete in ways we cannot now detect. This book argues from the pattern of what is present: the preserved disagreements, the named critics, the coexisting schools, not from arguments from silence. That is the strongest ground available. It is not the only ground possible. Structural claims of this kind rest on patterns across the record, not on single documents, and the difference is deliberate: the two kinds of claim have different logical structures and require different standards.

What would falsify it? A tradition with concentrated authority and strong absorptive mechanisms that produced rupture in the absence of forcing conditions would be a counter-example. So would a tradition with all the forcing conditions — distributed authority, framework competition, precision tools surfacing undismissable residuals — that did not produce rupture. These cases are not easy to find, but they should be possible to construct in principle. A framework that cannot be falsified even in principle is not a framework. It is a description that has stopped doing the work of explaining and started confirming itself.

Naming the limits does not weaken the central claim. It bounds it. A framework that explains everything explains nothing. A framework that explains some things, under specifiable conditions, against the boundary of what it does not explain, is the kind of framework that can be tested. That is what this one is offered as.

Kuhn describes rupture. Lakatos describes resistance. Taleb describes break. None specifies when rupture becomes unavoidable. This framework defines that condition: when an anomaly becomes visible, persistent, and inescapable, within a structure that cannot absorb it. Without that condition, these accounts describe outcomes without explaining their necessity.

The pattern did not end with Kepler. Newton's system, which unified celestial and terrestrial mechanics, held for more than two centuries, until a small discrepancy in Mercury's orbit refused to be absorbed. The precession of Mercury's perihelion differed from Newtonian prediction by a few arcseconds per century, a residual so small it was almost invisible, but one that persisted through every attempt to explain it away. It required Einstein's general theory of relativity to resolve it. The structure repeats: precision reveals error, error becomes verdict, and the framework breaks. The forcing anomaly is always small. That is why it is dangerous.

The Kerala–Kepler comparison is precisely that test of the framework. Both traditions had anomalies. What they did not share was a forcing anomaly, an error too precise and too stubborn to absorb without confronting the model itself. Kerala's computational inconsistencies were solvable within the existing framework, and were solved. Kepler's Mars residual was not. The mathematical sophistication was comparable. What differed were the conditions: the precision instruments that made the residual undeniable, the institutional structure that made it consequential, and the absence of an absorptive mechanism sufficient to contain the pressure. If those conditions are not doing the explanatory work — if the framework is wrong — the comparison should show no pattern. It shows a pattern.

The question the book leaves open is not only about Indian astronomy. It is about any system of knowledge that is working well: what is the error it cannot yet see, and what would it take to make that error impossible to ignore?

In every case, the answer did not come from success. It came from the point where success was no longer enough.

The Same Question

The second part asks the same question in a different domain. Not a historical system, but a current one. What conditions allow a working system to see its own failures? What does it take for an anomaly — visible, persistent, and no longer absorbable — to become a verdict?

The first part was not background. It was the control. What follows is the experiment.

Part Two

How a system stays there

Chapter One

The Question the System Did Not Ask

A few years ago, I developed a back problem.

I am describing it this way (a back problem) because that is all I ever learned about it. Not what caused it, not what was happening in the underlying tissue, not what the mechanism was or what risk it represented or why it resolved. A back problem. Something hurt. Then it did not. The system processed me. The file closed.

I want to trace what happened, precisely, because the precision matters. I am not telling this story to complain about my care. The people who treated me were not negligent. They followed reasonable protocols. I am telling it because I am a trained scientist, and somewhere in the middle of the sequence of events I am about to describe, I realized that the thing that was bothering me was not the pain. It was the absence of a question.

♦

The first visit: I described the symptom to my primary care physician. She recommended an X-ray. Reasonable. Rule out the obvious structural problems. The X-ray showed nothing significant.

The second visit: Physical therapy. Also reasonable. Back pain resolves with physical therapy more often than with intervention. I did the exercises. The pain persisted.

The third visit: More physical therapy. I followed the protocol. No meaningful relief.

The fourth visit, some months later: an MRI was ordered. Now we were going deeper. Good. The MRI was done. Reading it required a specialist. A referral was made.

The fifth visit: the specialist. To discuss the MRI, a fresh X-ray was required. Not the one taken months earlier. A new one. The specialist would not see me otherwise. I did not go quietly. I went back home and faced the wrath of my wife, who had a simple and entirely rational argument: the insurance covers it, the doctor ordered it, follow the order. For her, this was not complicated. For me, logically, it did not make sense. An X-ray already existed. The MRI — the more informative image, the one that had prompted the referral — was already in the system. Why was a new X-ray the price of entry to the conversation about the MRI? I did not have a good answer. I still do not. What I eventually understood was that the earlier image was not clinically insufficient. It was simply not recent. The system had a requirement. The requirement generated a procedure. The procedure generated a bill. Whether it generated information is a different question, one the system was not designed to ask.

The sixth visit: the results. The specialist reviewed the imaging. There was, she explained, nothing that required intervention. And then I said something that ended the encounter: I mentioned that I had been feeling better lately. The symptoms had largely resolved on their own.

She nodded. Come back when the symptoms return.

That was it. The encounter ended. I walked out.

♦

I spent years in a molecular biology laboratory learning that a result is not a finding until you understand what produced it. A cell that behaves unexpectedly is not a nuisance to be managed. It is a question. The unexpected behavior is the data. The data is what you follow. You do not close the file when the anomaly resolves. You ask what the resolution tells you about the mechanism.

The symptom resolved because I said so. I reported improvement. The system accepted the report and closed the encounter. There was no independent verification of whether anything had actually changed in the underlying tissue. There was no risk assessment of what might have caused the symptoms, whether that cause persisted, and what the probability was of recurrence. The patient reported feeling better. The system had no reason to pursue the question further.

And here is the thing that stopped me, as I walked to my car after that sixth visit. I had comprehensive medical insurance. One hundred percent coverage. My out-of-pocket cost for each of those six visits was fifteen dollars. Fifteen dollars. Less than a lunch. The financial barrier between me and the full resources of the system was so low as to be nearly invisible. No deductibles, no prior authorization fights I could not navigate, no point at which the cost of the next visit became a reason not to go. I had access to a primary care physician, a physical therapist, an imaging center, and a specialist. I had the educational background to describe my symptoms precisely, to understand the explanations I was given, and to ask follow-up questions when something was unclear. I had every advantage the system offers.

And the question was still not asked.

That absence did not feel like an error. It felt normal. That was the part I could not ignore.

♦

I started thinking about the person who does not have those advantages.

My cost was fifteen dollars a visit. Now consider the person for whom the number is not fifteen dollars but one hundred and fifty, or three hundred, or a day's wages. The person who skips the third visit because the copay requires a calculation. The person who never gets the MRI because the insurance authorization is denied or delayed until the symptom resolves on its own and the case is closed. The person who goes back to work with the pain because missing another day is not possible. That person does not receive the incomplete care I received. They receive an earlier termination of a process that was already not asking the right question. The system failed to understand my back problem at fifteen dollars a visit. What does it do at three hundred?

This is not a story about bad doctors or broken insurance or a uniquely dysfunctional corner of the healthcare system. This is a story about what the system is designed to do. And what the system is designed to do is process the presenting symptom. Not understand it.

♦

I do not know what caused my back problem. I do not know whether the resolution was genuine healing, temporary remission, or simply my body adapting to a chronic condition in a way that made the pain less salient. I do not know whether the imaging showed something real or an artifact, whether the physical therapy did anything useful or whether the improvement would have happened anyway, whether there is a structural finding that matters or whether the whole episode was a biological noise event that resolved on its own timeline regardless of anything the system did.

This uncertainty is real. It is not an excuse. It is part of the argument.

The question this book asks is not whether medicine learns. It asks what the conditions are under which the learning becomes mandatory rather than optional, and what happens to the anomalies that arrive before those conditions are in place. The argument is not that medicine is failing. It is that the conditions for mandatory learning are present in some domains and absent in most, and that the pattern of their absence is structural, not accidental.

One of the structural features of medicine that makes systematic learning difficult is that biology is genuinely noisy. Outcomes are uncertain. The same symptom in two patients may have two completely different causes, two completely different natural histories, and two completely different responses to the same treatment. This noise is real, and it makes the attribution of outcomes to specific interventions genuinely difficult. Medicine is not engineering. A physician cannot test the patient the way an engineer tests a bridge.

But the noise also provides cover. In a system where outcomes are uncertain and attribution is difficult, the failure to ask the underlying question is invisible. Nobody died because my back problem was processed rather than investigated. The system has no mechanism for counting the cost of the question that was not asked: the diagnostic understanding that was not reached, the risk assessment that was not performed, the early warning signs that were not identified. These are not events. They are absences. Absences do not generate incident reports.

And so the system continues, processing presenting symptoms, closing files when symptoms resolve, sending patients home when they report improvement, not because this is the best it can do, but because the structure does not require anything more. What is not incentivized does not, systematically, happen.

♦

The first part of this volume examined exactly this: a tradition that had the tools, the data, and the individual brilliance — and did not produce the next step, because the error in the existing framework was never forced into the open.

In astronomy, the anomaly was an eight-arcminute discrepancy in the predicted orbit of Mars. Kepler refused to absorb it. The system broke. In medicine, the anomalies are everywhere. They are in the unexplained symptom that resolves without a clear intervention. They are in the patient who does not respond to a treatment that works for most patients. They are in the mortality rate that is twice as high in one hospital as in an otherwise comparable one. They are visible. They are not being absorbed into understanding. They are being absorbed into the background noise of a system that does not have a structural requirement to ask why.

This book is an attempt to understand that structure. Not to accuse medicine of bad faith. Not to suggest that the physicians who treated my back problem were incompetent or careless. The system is genuinely complex, the biology is genuinely noisy, and the humility I brought to the astronomy book I am trying to bring here as well.

But humility about what I do not know is not the same as silence about what I can see. And what I can see — what my training as a scientist and my daily exposure as a patient and an observer has made increasingly difficult to unsee — is a system that is better designed to sustain its existing practice than to question it. A system in which the filters on knowledge pass favorable findings more easily than unfavorable ones. A system in which the profession evaluates its own performance by standards it sets for itself. A system in which the error that matters most is the error that was never identified as an error, because the question that would have identified it was never asked.

In astronomy, the question was why a system did not produce the breakthrough its evidence supported. In medicine, I am asking why a current system does not learn from its failures at the rate that its patients' lives require.

The question is the same. The stakes, for the people being processed by the system right now, are higher.

♦

Even if I had gone back — even if I had returned to the clinic and said: I am still concerned, I want to understand what happened, I want a risk assessment, I want someone to tell me what to watch for — what would the system have done? It would have ordered another test. And if the test showed nothing actionable, it would have sent me home again. Because the system has two modes: manage the symptom, or operate on the structure. It does not have a third mode, which is understand what happened and why.

The third mode is not a medical procedure. It is a conversation oriented toward mechanism rather than toward action. It is the physician asking: what do we think caused this, what does the resolution tell us, what is the probability of recurrence, what would recurrence look like and at what point would it warrant a different response? These are the questions a scientist asks when an experiment produces an unexpected result. They are also the questions a patient deserves to have asked on their behalf.

They are not asked, systematically, because they are not billable. They do not produce a test result that can be documented in the chart. They exist in the space between the presenting symptom and the next presenting symptom, and that space is not part of the system's architecture.

What the system cannot see, it cannot fix. What it cannot fix, it continues.

This is what a local maximum looks like from the inside. The system is not failing, not by any measure it applies to itself. It is processing cases, producing diagnoses, generating treatment plans. Most patients improve, or at least do not obviously worsen. The residuals that do not fit are absorbed into the noise of biological complexity or deferred to the category of cases that require more information. The system continues because it works well enough that the pressure to change never accumulates into something that cannot be managed. A local maximum is not a failure. It is a stable position from which the next step cannot be seen because the current position provides no reason to look for it.

I did not know it then, as I walked away from that sixth visit, that this was not an isolated experience. The absence I had noticed — the question that was never asked — had appeared before, under conditions where the stakes were not discomfort but death.

That is where this book begins.

Chapter Two

The Ward Where They Died

In the spring of 1847, a young Hungarian physician named Ignaz Semmelweis noticed something that should not have been possible to ignore.

Vienna General Hospital ran two obstetric wards side by side. The First Division was staffed by medical students and doctors. The Second Division was staffed by midwives. Women admitted to the First Division died of childbed fever at a rate of roughly ten percent. In the Second Division, the rate was less than four percent. The difference was not subtle. It was not a statistical artifact. It was the kind of difference that fills a ward with bodies.

Women knew about it. Patients admitted to the hospital would beg to be placed in the Second Division. Some gave birth in the street on purpose, preferring to deliver alone on the cobblestones rather than risk the First Division. The mortality gap was a matter of street-level knowledge among the women of Vienna before it became a subject of medical investigation.

Semmelweis noticed. He began to ask why.

The conventional explanation was miasma: bad air, foul vapors rising from the city, invisible and incurable. It was the all-purpose explanation of the era, useful precisely because it explained nothing in particular and therefore could not be tested. Semmelweis did not find it convincing. The two wards shared the same air. They were separated by a wall, not a weather system.

Then, in the spring of that year, a colleague named Jakob Kolletschka died. He had been performing an autopsy when a student's scalpel slipped and cut his finger. The wound became infected. The infection spread. Kolletschka died of the same constellation of symptoms — the same fever, the same organ failure, the same progression — as the women dying in the First Division.

Semmelweis saw the connection immediately. Medical students and doctors came to the obstetric ward directly from the autopsy room. They had been handling cadavers. They carried something on their hands (some material from the dead) that they then brought into contact with women in labor. The midwives in the Second Division performed no autopsies. They brought nothing from the dead.

He introduced a simple intervention: physicians and students were required to wash their hands in a chlorinated lime solution before examining patients. The results were immediate. Mortality in the First Division fell from roughly ten percent to less than two percent. The intervention worked. The anomaly was explained. The data was unambiguous.

At that point, in any system designed to learn, the question would have been settled.

It was not.

The medical establishment did not accept it.

♦

What happened next is one of the most studied, most cited, and most uncomfortable episodes in the history of medicine. Semmelweis was not ignored: he was actively resisted. Senior physicians disputed his findings. Colleagues wrote papers arguing against him. The germ theory of disease had not yet been formulated, and without it, the mechanism Semmelweis proposed (invisible particles from cadavers causing fever and death) struck many as implausible, even offensive.

There was also a professional dimension to the resistance. To accept Semmelweis's conclusion was to accept that doctors were killing patients. That the very act of medical examination — the physician's trained hands, the physician's learned touch — was a vector of death. This was not merely a scientific claim. It was an accusation. And the profession, which controlled the terms of its own evaluation, was not inclined to accept an accusation from a junior Hungarian physician with a peculiar theory about handwashing.

Semmelweis grew increasingly agitated as the years passed and the deaths continued. He wrote obsessive letters. He published his findings in a book that was poorly organized and nearly unreadable in its intensity. He descended, ultimately, into what appears to have been a breakdown. In 1865, he was committed to a mental institution, where he died two weeks later, ironically, of an infection. He was forty-seven years old.

Routine handwashing in obstetric wards did not become standard practice until the 1880s, after Louis Pasteur and Joseph Lister had provided the germ theory framework that made Semmelweis's observations comprehensible. The evidence had been present for thirty years. The intervention was simple. The deaths were preventable. They continued anyway.

♦

The Semmelweis case is often told as a story about individual resistance, about the pride and defensiveness of specific physicians who could not accept an uncomfortable truth. That reading is not wrong, but it is incomplete. What happened in Vienna was not only a failure of individuals. It was a failure of structure.

Medicine in the 1840s had no mechanism for forcing its own errors to become visible. There was no independent body to investigate the mortality gap. There was no mandatory reporting of outcomes that would have made the First Division's death rate a matter of institutional accountability. There was no feedback loop that connected the pattern in the data to the behavior that was causing it.

The anomaly was visible to anyone who looked. The system was not built to look.

There is still no independent body that investigates medical adverse events the way aviation investigates its accidents: routinely, nationally, with public findings and mandatory system response. Outcome reporting is still largely voluntary and largely internal. The anomaly is still visible to anyone who looks carefully. The system is still not built, structurally, to require that anyone look.

This is not 1847. The most visible failures of Semmelweis's era are gone. The structural absence at the center of that story has not been solved.

My back problem is not in the same category as the deaths in Semmelweis's ward. I am not making that comparison. What I am saying is that the structural question is identical: what would it take for the system to treat the unexplained as a question it is obligated to pursue, rather than a nuisance it is permitted to absorb?

In Vienna in 1847, the answer was: thirty years and a new theory. In my case, the answer was: come back when the symptoms return.

Both answers describe a system that is waiting for the anomaly to become undeniable before it asks why. The cost of that waiting is paid by the patients who are inside the system while it waits.

Chapter Three

When the Evidence Arrives Too Early

Medicine is not the only system that defers evidence it finds inconvenient.

Before we can follow what happened to inconvenient evidence, we need to understand how a system decides what counts as evidence, and what it is permitted to ignore.

I spent years as a postdoctoral researcher in a biomedical laboratory. During that time I watched a structural fact about the academic system operate in plain sight: too many PhDs were being trained for faculty positions that did not exist in sufficient numbers. The postdoc's primary function in the laboratory economy was skilled labor at wages that no other comparably trained professional would accept. Every principal investigator knew this. Every department chair knew this. The data had been in the literature for decades. The pattern was visible, documented, and widely acknowledged whenever it became impossible to ignore.

And then each laboratory continued exactly as before. The acknowledgment cost nothing. The absorption required nothing to change. The framework for acting on it — the incentive structure that would have required a laboratory to take fewer postdocs, or to be honest with them about their prospects, or to invest in their careers rather than their projects — belonged to no one in particular. Everyone was responsible. No one was accountable. The system continued.

♦

The same pattern operates in medicine, and the clearest historical demonstration does not involve a controversial claim. It involves citrus fruit.

Scurvy was one of the great catastrophes of naval medicine. For centuries, sailors on long voyages died of it in numbers that would be inconceivable today. The disease was not subtle. It began with fatigue and joint pain, progressed to bleeding gums and loosening teeth, and ended, in its advanced stages, in hemorrhage, organ failure, and death. It was the leading cause of mortality on extended sea voyages. It killed more British sailors in the eighteenth century than combat did.

The cause was a deficiency of vitamin C, though no one knew that yet. What was known, empirically, was that sailors who ate fresh food did not develop scurvy, and sailors who subsisted on salt meat and ship's biscuit did. This observation had been made by ship's surgeons for generations. It was not a secret. It was not a minority view. It was a pattern visible to anyone who had spent time at sea.

In 1747, a Scottish naval surgeon named James Lind decided to test the pattern systematically. He took twelve sailors suffering from scurvy, divided them into six pairs, and gave each pair a different treatment while keeping everything else as constant as the conditions of an eighteenth-century warship permitted. The treatments included cider, diluted sulfuric acid, vinegar, seawater, a mixture of nutmeg and other compounds, and — for one pair — two oranges and a lemon daily. Within a week, the sailors who received the citrus were largely recovered. The others were not.

Lind published his findings in 1753 in a treatise on scurvy. The evidence was clear. The trial was not perfect by modern standards — twelve patients, one comparison, no randomization in the contemporary sense — but it was more systematic than almost anything else being done in medicine at the time. The citrus intervention worked. Lind had shown it.

The British Navy did not issue citrus juice as a standard antiscorbutic ration until 1795, forty-two years after Lind's publication. By conservative estimates, tens of thousands of sailors died of scurvy in those intervening decades. The evidence had arrived. The action did not follow.

The question that matters is not why the Navy was slow. The question is what kind of slowness this was.

One answer is bureaucratic inertia: large institutions change slowly, procurement systems are rigid, the men who made decisions about naval provisioning were not the men who watched sailors die. There is truth in this. Bureaucratic inertia is real and it contributed to the delay.

But there is a more precise answer. Lind knew that citrus prevented scurvy. He did not know why. The concept of vitamins did not exist. The biochemical mechanism by which ascorbic acid prevents the breakdown of collagen — which is what scurvy is, at the cellular level — would not be understood for another century and a half. Without the mechanism, the finding existed in an explanatory vacuum. And medicine — then as now — is systematically more comfortable with interventions that fit within a coherent explanatory framework than with interventions that do not.

A finding without a mechanism is harder to generalize, harder to extend to other patients and other conditions, and considerably easier to argue against. Perhaps the sailors who recovered after eating citrus were already closer to recovery regardless of the treatment. Perhaps there was some confounding factor — the fresh food more broadly, the improvement in morale, the change in diet — that accounted for the improvement. Perhaps Lind's sample was too small to draw firm conclusions. Without a mechanism, these objections cannot be fully refuted. They are not good objections — the pattern was too consistent and too large to be explained away — but they are defensible objections. And in a system where defensible objections are sufficient to defer action, they were sufficient.

The cost of waiting for the mechanism was paid by the sailors who died of scurvy between 1753 and 1795. They are not individually remembered. They are a number in an estimate. They died of a disease that was, by the time of their deaths, preventable by a treatment that was available, documented, and supported by evidence. They died because the evidence had not yet been given a framework within which the system was required to act on it.

♦

Now consider a case where the mechanism was not absent. Where the researcher knew exactly why the treatment worked, had the experimental evidence, and still waited thirteen years.

In the early 1980s, two Australians, a physician named Barry Marshall and a pathologist named Robin Warren, identified a bacterium (*Helicobacter pylori*) in the stomachs of patients with gastritis and peptic ulcer disease. They cultured it, treated patients with antibiotics, and watched ulcers that had been considered chronic and incurable resolve completely. They had the evidence. They had the mechanism. Ulcers were caused by a bacterium, could be cured with a two-week course of antibiotics, and had nothing to do with the stress and excess acid that medicine had attributed them to for decades. The journals rejected their papers. Conference presentations were met with skepticism. Senior gastroenterologists, whose entire practices were built around the lifelong management of a chronic condition, were not persuaded. The stress hypothesis had been dominant for so long that the alternative felt not just wrong but absurd: a bacterium, in a stomach environment everyone knew was too acidic to support bacterial life.

In 1984, Marshall drank a Petri dish containing a culture of *H. pylori*. Within days he had gastritis. He took antibiotics and recovered. He had given himself the disease and cured it. The act was not primarily a scientific proof: a single self-experiment cannot constitute one. It was a forcing act. It made the anomaly impossible to contain. Here was a physician who had deliberately given himself a disease the establishment said did not work the way he claimed, and had then cured it the way the establishment said was impossible. The anomaly escaped the filters. It became something that could be told in plain language to people who were not gastroenterologists.

From the initial findings in the early 1980s to the NIH consensus statement recommending antibiotic treatment in 1994 was roughly thirteen years. During those thirteen years, patients with peptic ulcers were told they had a chronic condition requiring lifelong management. They were prescribed antacids. They were advised to reduce their stress. They were not told they had a bacterial infection curable in two weeks, because that knowledge, though available, had not yet been accepted by the system. Marshall and Warren received the Nobel Prize in 2005. The recognition, when it came, was unambiguous. The cost of the delay — how many unnecessary surgeries, how many decades of antacids, how many complications from ulcers that could have been cured — was never calculated. Medicine does not routinely tabulate the cost of delayed adoption. The delay is absorbed into the background. The patients inside it are not counted.

The contrast with Lind is instructive. Lind had the pattern and not the mechanism. Marshall had both. The absence of mechanism explains the forty-two years. The presence of mechanism does not explain the thirteen. Something else was required to defer the Marshall evidence for thirteen years, something that the mechanism alone could not overcome. That something is the structure of how medical knowledge is produced, filtered, and accepted. It is the same structure in both cases. The mechanism merely changed how long the deferral could be sustained.

♦

The deeper structure is this: the evidence for citrus arrived before the framework that would have made acting on it mandatory. The framework, in this case, was not just biochemical knowledge. It was also institutional. A framework for action requires not only an explanation of why something works, but a mechanism that makes acting on the evidence the default rather than the exception, that makes not acting require justification rather than acting requiring justification.

Medicine in the eighteenth century had no such mechanism. Neither, in most domains, does medicine today.

Consider the structure of how medical evidence becomes practice. A finding is published. It is reviewed by other researchers. If it survives review, it enters the literature. From the literature, it may eventually be incorporated into clinical guidelines, a process that typically takes years, requires a threshold of replication that is rarely precisely defined, and is managed by expert panels whose composition and processes vary widely. From the guidelines, it may eventually reach practicing physicians, through continuing education, professional society communications, and the ordinary processes by which practitioners keep up with their field.

At each step, the evidence must pass through a filter. At each step, the filter is easier to pass in one direction than the other. Evidence that supports an existing practice moves through the system with less friction than evidence that challenges one. Evidence that fits an established theoretical framework is more readily accepted than evidence that does not. Evidence that implies no change in behavior is less threatening than evidence that implies a significant one. The filters are not designed to be asymmetric. They are asymmetric because the people who control them are embedded in the existing system and have interests in its continuation.

Lind's citrus evidence was not suppressed. It was deferred. That is a crucial distinction. Suppression is active. Deferral is passive. It requires nothing of anyone. It simply requires that the system continue doing what it was already doing while the evidence waits for a framework strong enough to force a change.

♦

The question this chapter asks is not whether caution is appropriate. It is. The question is whether the caution is symmetric.

When a new drug is proposed for approval, the standard of evidence is high. Years of clinical trials. Multiple phases of testing. Replication across populations. The system is designed to be cautious about adopting something new, and the caution is generally justified: new treatments can cause harm, and the history of medicine includes enough examples of harmful treatments that were prematurely adopted to make the caution rational.

But what is the standard of evidence for abandoning an established practice? For revising a guideline that has been in place for a decade? For acknowledging that a treatment widely used for a common condition has been shown, in multiple replicated trials, to be no more effective than placebo?

The standard is not symmetric. The threshold for adoption is high and relatively clearly defined. The threshold for abandonment is higher, less clearly defined, and more easily deferred. It is a structural feature of how the filters work, not a formal policy, but an emergent consequence of who controls the process. Evidence accumulates against a practice. The practice continues. The guideline remains. The deferral is individually defensible at every step — one more study, one more replication, one more consensus panel — while the patients who are being treated under the existing guideline absorb the cost of the delay.

The sailors who died of scurvy between 1753 and 1795 were not killed by malice. They were killed by a structure that required more than the evidence could provide before it was willing to act. That structure has been modified in the centuries since. It has not been resolved.

♦

The forty-two year delay between Lind's publication and the Navy's adoption of citrus juice was not simply a failure of knowledge. It was also a failure of incentive. The people who suffered from scurvy were sailors. The people who made decisions about naval provisioning were administrators and officers. The cost of the delay was borne by one group. The decision to end the delay required action by another group. And that second group had no structural obligation to act on behalf of the first.

This is the pattern I watched in the postdoc system. The people who bore the cost of the structural problem — the postdocs working at wages that did not reflect their training, preparing for careers that largely did not exist — were not in a position to change the system. The people who were in a position to change it — the principal investigators, the department chairs, the funding agencies — bore none of the cost themselves. Their laboratories ran efficiently. Their grants were funded. Their papers were published. The incentive to change was absent because the cost of not changing was not theirs to pay.

The asymmetry is not an accident. It is produced by the distribution of costs and benefits in the system. The evidence tends to be deferred longest when the cost of acting on it falls on people who cannot require action, and the benefit of deferral falls on people who can. Change the distribution and you change the incentive to defer. That is a structural change. It is harder than publishing a guideline. It is the only change that would actually work.

Chapter Four

The Noise That Hides the Signal

I said I was feeling better, and the file closed. But I did not actually know if I was better.

I mean this precisely. I knew the pain had diminished. I did not know whether the underlying cause — whatever it was — had resolved, or was temporarily quiet, or had never been what was generating the pain in the first place. I did not know whether the physical therapy had done something useful or whether the improvement would have happened on the same timeline without it. I did not know whether the imaging had shown something real or a finding so common in asymptomatic people that it was essentially noise. I reported a subjective improvement. The system accepted the report. That is not the same as knowing what happened.

The system has no mechanism for distinguishing "patient reports improvement" from "patient is better." In most cases, these are the same thing. In some cases, they are not. The system cannot tell the difference, and is not currently designed to try.

And when a system cannot distinguish signal from noise, it defaults to treating everything as noise.

Once that default sets in, nothing in the system requires it to change.

This is the noise problem. And it runs deeper than my back, and deeper than any individual encounter. It is built into the biology.

♦

When an engineer tests a bridge, the system behaves in predictable ways. Apply a known force; measure a known deflection. The relationship between cause and effect is tight. Variation is small. The background is quiet enough that signals are easy to detect against it.

When a physician treats a patient, the system is different in kind, not just in degree. Two patients with the same diagnosis, treated identically, may have dramatically different outcomes. One recovers; the other does not. The reasons may be genetic, immunological, environmental, or simply random: stochastic variation in biological processes that are too complex, too individualized, and too incompletely understood to fully model. Medicine calls this patient heterogeneity. It is real. It is pervasive. And it makes the detection of signal against noise genuinely, not rhetorically, difficult.

A molecular biologist learns this early. In the laboratory, a cell that behaves unexpectedly is a signal. But before you can call it a signal, you have to rule out the noise: the contamination, the pipetting error, the batch effect, the variation between cell lines, the day-of-week effect on assay sensitivity. Most unexpected results, examined carefully, turn out to be noise. The discipline of experimental biology is largely the discipline of learning to distinguish signal from noise before drawing conclusions.

Medicine faces the same problem, at far greater scale, in far messier systems, with far higher stakes for being wrong in either direction. This is why the randomized controlled trial was invented. By randomly assigning patients to treatment and control groups, a sufficiently large trial averages out the heterogeneity. The individual variation cancels. The effect of the treatment — if there is one — becomes visible against the background. The randomized trial is medicine's version of the controlled experiment: a tool for finding signal in a system where noise is the default condition.

It is a powerful tool. It is not the whole story.

♦

Consider precisely what a randomized trial measures. It measures the average effect of a treatment across a defined population. If a treatment benefits sixty percent of patients and harms forty percent, and the average benefit outweighs the average harm, the trial reports a positive result. The drug is approved. The guideline recommends it.

The forty percent who were harmed are real people, but they are invisible in the aggregate number. They have been averaged out. Their signal has been absorbed into the noise of the trial.

The question is whether the information that gets averaged out — the pattern of who is harmed, under what conditions, with what characteristics — is being systematically captured and acted upon somewhere else in the system.

In a well-functioning system, the answer would be yes. The patients who did not respond would be studied. The adverse events would be tracked and used to refine the understanding of when the treatment is appropriate and when it is not. Medicine has mechanisms for this — pharmacovigilance systems, adverse event reporting, post-market surveillance — and they are incomplete, inconsistently applied, and systematically underfunded compared to the mechanisms for establishing efficacy in the first place.

The result is a system that is substantially better at detecting that a treatment works, on average, than at detecting that it works harmfully for a specific subgroup, or that it stops working over time, or that the population it was tested on differs in important ways from the population now receiving it. The noise absorbs these signals. The anomaly stays contained. The trial was positive; the drug is approved; the cases that do not fit are individual variation.

♦

There is a second way that noise hides signal in medicine, and it operates not in the statistics but in the language.

When a patient dies unexpectedly after surgery, the default attribution is the underlying disease. The cancer was advanced. The heart was weak. The patient was never a strong candidate for the procedure. When a patient develops a complication after an intervention, the complication is attributed to the complexity of the case. When a patient does not respond to a treatment that works for most patients, the patient is considered to have had a more aggressive form of the disease. The language of medicine is full of constructions that locate the source of bad outcomes in the patient rather than in the system.

The attribution to patient factors is frequently accurate. But *frequently accurate* is not the same as *systematically examined*.

The question is not whether individual attributions are sometimes correct. The question is whether the system has a mechanism for distinguishing the cases where the outcome was genuinely driven by the underlying disease from the cases where it was driven by the treatment, the decision, or the process. In most hospitals, for most adverse events, the answer is no. The attribution to patient factors is the default. Departing from the default — saying that the outcome was produced by the system rather than the disease — requires active investigation, carries professional and legal risk, and happens far less often than the frequency of system-driven adverse events would warrant.

The noise provides the cover. Every bad outcome in medicine can, plausibly, be attributed to the underlying biology. The biology is complex enough, variable enough, and poorly enough understood that the attribution is rarely obviously wrong. And in a system where plausible attribution is sufficient to close the file, the signal — the genuine system failure, the preventable error, the treatment that was doing harm — stays hidden in the noise.

♦

Here is the contrast that clarifies the structural problem.

Aviation also operates in a noisy environment. Mechanical systems fail unpredictably. Weather is variable. Human performance is inconsistent under stress. And yet commercial aviation has become, over the past fifty years, dramatically safer — the fatality rate per passenger mile has fallen by something close to ninety-five percent — because aviation built a system that treats every serious anomaly as a signal, regardless of how plausible the noise explanation might be.

A human body is not a manufactured system. We do not have the original blueprints, no two are built identically, and the failure modes were not documented by an engineer. The aviation analogy has limits, and the medical community knows them intimately. But the question the analogy raises is not whether medicine is as predictable as aviation. It is whether medicine has built the architecture for learning from failure that aviation built precisely because of its unpredictability, and whether the absence of that architecture in medicine is a fact of biology or a choice of structure.

When a commercial aircraft is involved in a serious incident, the investigation does not begin by asking whether the outcome can be attributed to weather, or mechanical complexity, or the difficulty of the approach. It begins by asking what happened and what the system can learn from it. The investigation is independent of the airline. The findings are public. The recommendations require formal response. The default is signal, not noise.

Medicine's default is the opposite. The default is noise: patient complexity, biological variability, the inherent unpredictability of disease. This default is not unreasonable; the noise is real. But a system organized around noise as the default explanation for bad outcomes will systematically underdetect the signal. It will attribute to biology what should be attributed to process. It will close files that should stay open. It will continue practices that should be questioned, because the questions require departing from the default, and the default is comfortable, and the structure does not require anyone to be uncomfortable.

My symptom resolved. The file closed. Was I better? I do not know. The system did not need to know. That is the noise problem, not that the biology is complex, but that the complexity of the biology is used as a reason not to ask. That is a question about filters.

Chapter Five

Who Gets to Decide What Counts

The specialist who would not see me without a new X-ray was not making a clinical decision. She was enforcing a filter.

The filter was not labeled as such. It was labeled as a requirement: a procedural step, a standard of care, a condition of the referral. But what it did, structurally, was determine what information entered the encounter and what did not. The MRI I had already taken — the more detailed, more recent, more informative image — was in the system. It was not sufficient. A new X-ray was required. The requirement generated a procedure. The procedure generated a bill. The information it generated, if any, was not the point. The filter had its own logic, and the logic was not primarily clinical.

I am using this example not to revisit my back problem but because it illustrates something precise about how medical knowledge works. Knowledge does not flow freely through medicine. It passes through filters at every level: from the research laboratory to the published literature, from the published literature to the clinical guideline, from the guideline to the individual encounter. Each filter has its own logic. Each is controlled by someone. And the someone who controls a filter always has interests that are not identical to the interests of the patient on the other side of it. These filters are the mechanism that determines whether the anomalies described in earlier chapters — the question not asked, the evidence deferred, the signal hidden in noise — become visible or disappear.

♦

The primary filter in the production of medical knowledge is the peer-reviewed journal. A finding that is not published in a peer-reviewed journal does not, for most practical purposes, exist within the medical knowledge system. It cannot be cited in guidelines. It cannot form the basis for treatment recommendations. It is not part of the evidence base from which physicians draw. Publication is not a step in the dissemination of knowledge. It is the gatekeeping function that determines what becomes knowledge in the first place.

Peer reviewers are experts in the field they are reviewing. That is their qualification, and it is genuine. Expert review does important work: it catches errors, identifies methodological weaknesses, and enforces minimum standards of rigor. But experts in a field are also invested in the field's existing understanding. They have built careers on the current framework. They have published papers that assume the current framework is correct. A finding that challenges the framework is not only an intellectual challenge to a peer reviewer. It is, in a small but real way, a challenge to the peer reviewer's own work.

It makes the peer review system structurally less efficient at passing findings that challenge established understanding than findings that extend it. The distinction between corruption and structural bias is important. Corruption requires intent. Structural bias requires only incentive. The peer reviewer who rejects a paper challenging the prevailing model may be acting in complete good faith, applying rigorous standards, and still be functioning as a filter that protects the prevailing model from challenge more than the evidence warrants.

The result is a literature that is better at recording what the field already believes than at surfacing what the field might need to revise. Semmelweis could not get his findings accepted not because the evidence was weak — it was not — but because the evidence required the field to believe something the field's experts found implausible. Marshall's early papers were rejected not because the methodology was flawed but because the conclusion — that a bacterium caused ulcers in an environment everyone knew was too acidic for bacterial life — did not fit within the existing framework. The filter did what filters do. It passed what fit and held what did not.

Not by rejecting the evidence outright. By asking for one more reason to wait.

And then another.

♦

The second filter is the clinical guideline. Guidelines translate published evidence into actionable recommendations for practicing physicians. They are produced by expert panels — convened by professional societies, government health agencies, and academic medical centers — and they represent the operational level at which the evidence base becomes clinical practice. A physician who follows the guideline is practicing evidence-based medicine. That is the definition.

Guidelines have a momentum that is independent of the evidence. Once a recommendation enters a guideline, it tends to stay there until the evidence against it is not merely sufficient but overwhelming, and the process of revision — which is slow, committee-driven, and subject to the same structural biases as peer review — is completed. In the interim, the gap between what the evidence now supports and what the guideline recommends can be substantial. The guideline is being followed. The patients are receiving evidence-based care. The evidence has moved. The guideline has not caught up.

There is a legal dimension to this that reinforces the gap. A physician who departs from a guideline in response to emerging evidence is practicing outside the standard of care. If the departure results in harm, the guideline's absence provides no legal protection. The physician who follows the guideline, by contrast, has a defense regardless of whether the guideline reflects current evidence. The incentive structure is unambiguous: follow the guideline. The guideline is the filter. And the filter moves slowly by design, in ways that systematically favor established practice over emerging evidence.

The caution is not symmetric. The standard of evidence required to get a new recommendation into a guideline is substantially lower than the standard required to get an old recommendation out. Adoption is easier than abandonment. The filter is asymmetric in a direction that protects existing practice.

♦

The third filter is financial, and it operates at every level of the system simultaneously.

The development, testing, and approval of new treatments in medicine is almost entirely funded by the companies that will profit from those treatments if they are approved. The structure of pharmaceutical development has worked this way for decades. Its implications for what the published literature contains — and what it does not contain — are more thoroughly documented than almost any other distortion in medicine, and still insufficiently acted upon.

Pharmaceutical companies conduct many trials, but not all results become equally visible. Trials showing that a drug does not work, works no better than an existing cheaper treatment, or causes harm in a subpopulation are less likely to reach the literature physicians actually read. This is publication bias, and its scale is not trivial. A physician reading the published literature on a drug may be reading a systematically curated selection of the evidence, curated, in part, by the party with the greatest financial stake in a particular conclusion. In any other adversarial setting, we would recognize this as a disclosure problem. In medicine, it becomes the evidence base. The filter is not neutral. It is not designed to be.

The financial filter does not operate only at the level of pharmaceutical research. It operates at the level of the specialist who cannot see a patient without a new X-ray, at the level of the hospital whose revenue depends on procedure volume, at the level of the professional society whose conferences are funded by the industry whose products the society's guidelines evaluate. At each level, the financial interest is not identical to the clinical interest. At each level, the filter passes what is financially compatible and holds what is not.

♦

The three filters share a structural feature more important than their individual mechanisms: they are all controlled by parties whose interests are not primarily the surfacing of anomalies.

A system designed to surface anomalies would be organized differently. Its filters would be controlled, at least in part, by parties whose explicit function was to find what the system had missed: to pass precisely the findings that challenged established understanding, to treat anomalies as the most valuable signal in the data rather than the most threatening. Such a system does not currently exist in medicine at scale. It exists in fragments: in the investigative journalism that occasionally exposes an unpublished trial, in the independent researchers who make careers of questioning established treatments, in the rare regulatory action that forces a safety re-evaluation. These are not the primary mechanisms. They are the exceptions.

The most documented case of this failure is Vioxx.

Merck's painkiller was approved by the FDA in 1999. In 2000, Merck's VIGOR study — designed primarily to demonstrate the drug's gastrointestinal safety advantage — reported as a secondary observation that patients taking Vioxx had heart attacks at roughly fourfold to fivefold the rate of patients taking naproxen, the comparator. The signal was real but methodologically ambiguous: VIGOR's active-comparator design could not cleanly distinguish between Vioxx increasing cardiovascular risk and naproxen lowering it. Merck advanced the naproxen interpretation. A 2001 *JAMA* meta-analysis by Cleveland Clinic cardiologists Mukherjee, Nissen, and Topol advanced the harm interpretation. The scientific community did not reach consensus.

The drug remained on the market for about four years after the signal appeared, generating more than two billion dollars in annual sales. The FDA added cardiovascular information to the Vioxx label in 2002. The disambiguating evidence, when it came, came from another Merck-funded trial. APPROVe — a long-running, placebo-controlled study testing Vioxx for the prevention of recurrent colorectal polyps — supplied the control VIGOR had not. In September 2004, APPROVe's safety board stopped the trial early after finding that patients taking Vioxx had serious cardiovascular events at roughly twice the placebo rate. Merck withdrew Vioxx worldwide that September: five years after approval, four years after the original signal.

The accounting that followed framed the interval. FDA epidemiologist David Graham, testifying before the Senate Finance Committee in November 2004, estimated that Vioxx had been associated with between 88,000 and 139,000 excess serious coronary events in the United States, a substantial fraction of them fatal. In a 2005 Expression of Concern, the *New England Journal of Medicine* stated that data on three heart attacks in the rofecoxib group had not been included in the manuscript submitted to it; Merck disputed that characterization. Civil litigation produced documents indicating that company representatives had been directed not to raise cardiovascular findings with physicians who inquired. In 2007, Merck settled approximately 27,000 individual claims for roughly $4.85 billion.

The significance of Vioxx is not that a hidden signal was eventually exposed. The signal was visible. The significance is that visibility did not make it decisive. This is the financial filter operating exactly as the previous section described. The data were generated by the manufacturer. The decisive disambiguating trial was also a manufacturer trial. The system had no independent mechanism capable of settling the question while more than two billion dollars a year of revenue rested on the answer. The signal was published. The anomaly remained ambiguous. The verdict effectively had to wait for the manufacturer's next trial.

The response after Vioxx followed the same pattern. Congress strengthened the FDA's post-market surveillance authority, expanded trial-registration requirements, and supported active safety monitoring through large population databases. The system became better at detecting signals after approval. But the deeper architecture

— manufacturer-funded trials, financially interested trial design, and the absence of an independent mechanism to force rapid disambiguation when a profitable drug generated a contested safety signal — remained largely intact.

The primary mechanisms are the filters. And the filters, taken together, produce a system that is better designed to validate and disseminate established understanding than to challenge it. Anomalies that are compelling enough — that cannot be explained away, that generate enough public attention, that come with a mechanism that the field can accept — eventually pass through. Many do not. They are held at the filter, absorbed into the background, attributed to noise, and filed under complexity.

There is a deeper structural point here that connects this book to its predecessor. Sanskrit, at its most precise, encoded astronomical knowledge in a form so elegant and self-contained that it made certain questions effectively unaskable within the tradition, not because anyone forbade them, but because the format was not built to ask them. The billing code system does something structurally similar to medical encounters. It encodes what happened in a form that is procedurally complete and financially legible, while having no syntax for the question that was not asked. My six visits generated six procedure codes. Each code was correct. None of them had a field for mechanism. None had a field for risk. None had a field for the question the patient walked away still carrying. The system was not designed to record what it was not designed to ask. That is not a metaphor for bureaucratic failure. It is the architecture.

The new X-ray I did not understand as clinically necessary was not a failure of the system. It was the system working exactly as designed. That is the most important thing to understand about filters: they are not malfunctions. They are features. Understanding why the anomaly stays hidden requires understanding what the filters are designed to do, and for whom. But filters explain why anomalies are blocked. The next question is harder: under what conditions does one get through?

Before asking how an anomaly gets through, it is worth naming, plainly, what the system is actually optimizing for. Not what it says it optimizes for. Not what its practitioners intend. What the architecture rewards. A fee-for-service payment model rewards volume: doing more generates more revenue than doing less, regardless of whether the more produces better outcomes. A liability structure that punishes disclosure rewards concealment: an institution that acknowledges an error increases its exposure; one that attributes the outcome to patient complexity faces less. A guideline system centered in professional societies rewards consensus: heterodox findings must overcome the inertia of established practice before they can reach the standard of care. A publishing system that favors positive results rewards affirmation: studies that confirm existing treatment are easier to fund, easier to publish, and easier to cite than studies that challenge it.

Taken individually, each of these features has a partial justification. Taken together, they describe a system optimizing for throughput, defensibility, and institutional stability, not for truth, not for mechanism, and not for the patient who needs the question asked that the system is not designed to ask. The anomaly that the system cannot see is not hidden by accident. It is hidden by architecture.

Chapter Six

When Does an Anomaly Become a Verdict

The previous chapters have been circling a single question. Why does the anomaly stay hidden? The answer has come in layers: the question is not asked; the evidence is deferred; the signal is absorbed into noise; the filters pass what fits and hold what does not. Each layer is real. Each contributes to the result. But taken together, they describe a consequence without fully explaining the mechanism.

This chapter names the mechanism. It does so by asking a more precise version of the question: not why does the anomaly stay hidden, but what would it take for the anomaly to become impossible to ignore? What are the conditions under which a finding that does not fit the current framework cannot be deferred, absorbed, or explained away, and must, instead, force the framework to change?

The answer is three conditions. They are not complicated. They are, however, rarely simultaneous. And the rarity of their coincidence is the structural explanation for why most anomalies, in medicine and in other complex systems, do not force change.

♦

Systems do not change because anomalies exist.

They change when anomalies are made inescapable.

The first condition is that the anomaly must be visible, visible in a form that cannot be dismissed.

This sounds straightforward and is not. A single case is not a visible anomaly. It is an outlier, and outliers are the expected texture of a noisy biological system. A pattern in observational data is not a visible anomaly. It is a hypothesis, plausibly confounded by a dozen variables that were not controlled for. A randomized trial is visible, but a single trial can be dismissed as a fluke, a methodological artifact, or a finding that applies only to the specific population in the specific center where the trial was run.

Forcing visibility requires a specific combination: the anomaly must be large enough that chance cannot plausibly explain it, consistent enough across settings that confounding cannot plausibly absorb it, and clear enough in its implications that the inference from data to conclusion cannot be perpetually deferred. These three conditions within a condition are individually achievable. Achieving all three simultaneously, in a system whose filters are designed to find reasons to defer, is the difficulty.

Semmelweis had visibility. The mortality gap between the two wards was large, consistent, and clear. He could not get it accepted because the framework had ways to absorb his finding: attribution to other causes, demand for a mechanism that did not yet exist, deferral for further study. Visibility alone, it turned out, was not enough.

♦

The second condition is that the anomaly must be persistent: it must resist absorption into the existing framework.

Every framework has degrees of freedom. It can accommodate a certain amount of contradictory evidence without requiring fundamental revision. It can invoke confounding factors, patient heterogeneity, measurement error, the complexity of the underlying biology. It can call for more research. It can narrow the claim: perhaps the anomaly applies only to a subpopulation, only under specific conditions, only in the short term. These moves are not always dishonest. The biology is complex. Confounding is real. More research is often warranted.

But every framework has a capacity for accommodation that, once exceeded, forces revision. The question is how large that capacity is, and how the system determines when it has been exceeded. In a system designed to surface anomalies, the capacity is small and the threshold is explicit. In a system designed to absorb anomalies, the capacity is large and the threshold is implicit, shifting, and controlled by the people who benefit from not being required to respond.

Medicine's framework capacity is large, partly justified by genuine biological complexity, partly expanded deliberately, over decades, by the profession's insistence on controlling the criteria by which its own evidence is evaluated. The result is a framework that can accommodate an extraordinary amount of anomalous evidence before any individual actor is required to acknowledge that something needs to change.

♦

The third condition is the most socially demanding, and the one medicine most consistently fails to meet. The anomaly must be treated as a verdict — as evidence that the system failed and must change — rather than as a question that can be studied indefinitely.

In aviation, this condition is structural. When a commercial aircraft is involved in a serious incident, the investigation is not voluntary. It is conducted by an independent body — the National Transportation Safety Board in the United States, equivalent agencies elsewhere — whose sole function is to determine what happened and to issue safety recommendations for preventing it from happening again. The findings are public. The recommendations must be responded to, either by implementation or by formal justification for non-implementation. The system treats every serious anomaly as information it is obligated to use.

The investigation does not ask whether the outcome was probably caused by the underlying complexity of flight. It does not defer the findings pending further research. It produces a verdict. The verdict requires a response.

Medicine, in most of its domains, does not have this condition. A patient dies unexpectedly. If the death is reviewed at all, it is reviewed by a mortality and morbidity conference, voluntary in many institutions, confidential in most, and not connected to any mechanism that requires its findings to produce change. The review may identify that something went wrong. The identification may produce a conversation. The conversation may or may not produce a recommendation. The recommendation may or may not be implemented. There is no mechanism at any step that treats the anomaly as a verdict rather than a puzzle.

This is not an accident. It is the result of deliberate decisions, made over decades, by the medical profession and its legal and political allies, to keep the review process internal and protected. The reasoning offered was that external scrutiny would discourage honest self-evaluation. The consequence has been a system in which honest self-evaluation, when it occurs, is not required to produce change, and is never verified to have done so.

♦

Kepler's anomaly was eight arcminutes. Mars's predicted position, under the most refined version of the Ptolemaic model available to him, differed from its observed position by eight arcminutes, a discrepancy too small for naked-eye observation to detect, but not too small for Tycho Brahe's precision instruments. The anomaly was visible because the instruments made it so. It was persistent because every adjustment Kepler tried — every modified epicycle, every refined parameter — could not absorb it. And it became a verdict because Kepler refused to treat it as anything less. He wrote, famously, that he could not throw away those eight arcminutes, because they contained a proof of the reformation of all astronomy.

That third condition — the refusal to treat the anomaly as a nuisance to be managed rather than a verdict to be acted upon — was not structural. It was a choice Kepler made as an individual. It was a choice the institution of medicine, in most of its domains, has not made as a structure.

This is the asymmetry that matters. In astronomy, the forcing anomaly required an individual of unusual intellectual courage to treat eight arcminutes as a verdict rather than a rounding error. In aviation, the same function is performed by a structural requirement: the investigation is mandatory, the findings are public, the response is obligatory. The verdict does not depend on the courage of any individual. It is produced by the architecture.

Medicine has individuals of unusual courage, the Semmelweises and Marshalls who refuse to absorb the anomaly. What it does not have, at the structural level, is the architecture that makes the anomaly into a verdict regardless of whether anyone is courageous enough to demand it. The anomaly's fate — whether it forces change or disappears into the system — depends too heavily on the character of the individuals who encounter it, and too little on the design of the system they encounter it within.

♦

Three conditions. Visible, persistent, verdict. In practice, they are almost never simultaneous in medicine, not because any one of them is impossible to achieve, but because the system is not designed to require all three to be met before the anomaly is closed.

And when they do not coincide, the system continues.

An anomaly can be visible but absorbed by the framework's degrees of freedom: Semmelweis. It can be visible and persistent for decades but never treated as a verdict: scurvy. It can be treated by an individual as a verdict but fail to force systemic change because the system has no mechanism for converting individual conviction into institutional requirement: Marshall, waiting thirteen years.

Each failure is a different version of the same structural absence. The absence of a mechanism that makes the three conditions simultaneously mandatory is the reason the anomaly disappears. Not ignorance. Not malice. Architecture.

Chapter Seven

Competence Without a Counterfactual

In software engineering, when I write code, a colleague reviews it. A separate quality assurance (QA) group tests it against expected behavior, runs it against edge cases the original author did not anticipate. The review and testing are not formalities. They are structural requirements: the code does not ship without them. And the standard against which the code is reviewed is not the author's intention. It is behavior: does it do what it is supposed to do, under all the conditions it is likely to encounter? The gap between intention and behavior is precisely the thing the review and testing are designed to find.

This is the counterfactual. Not a philosophical abstraction: a practical mechanism. The code that shipped is compared to the code that should have shipped. The outcome that occurred is compared to the outcome that should have occurred. The gap between the two is the measure of performance. No gap, good work. Large gap, problem to be addressed. The accountability is structural because the comparison is structural.

I have spent years in medicine as a patient and an observer. In that time, I have never once been told how the care I received compared to the care I should have received. Not because the comparison is impossible to make — for many decisions, it is entirely possible — but because the system does not currently make it. The outcome is observed. The counterfactual is not. And in the absence of the counterfactual, the question that would make accountability meaningful — compared to what? — is never asked.

♦

There is a question that medicine almost never asks, and that every other performance-intensive field asks constantly: compared to what?

When a bridge engineer's design is evaluated, the evaluation is against a specification. When a software engineer's code is reviewed, the review is against behavior. When a pilot's performance is assessed, the assessment is against a standard that is independent of the pilot's own judgment about how well the flight went. In each case, the evaluation is comparative. It asks not just what happened, but what should have happened, and whether the gap between the two falls within acceptable bounds.

Medicine evaluates outcomes. But it rarely evaluates outcomes against what the outcome would have been under different treatment, because — with a very few exceptions — it cannot know. The patient who died after surgery: would she have died anyway? The patient who developed a complication: was the complication caused by the treatment, or would it have occurred regardless? The patient who did not respond to therapy: was this a failure of the therapy, or was the disease simply not responsive to any available intervention?

These questions cannot, in most cases, be answered at the level of the individual patient. There is no counterfactual. There is no parallel case in which the same patient was treated differently, allowing direct comparison. The outcome is observed. The cause is inferred. Inference, in a noisy biological system with a patient population too diverse to model fully, is always uncertain. This is a genuine epistemological difficulty, not a rhetorical one.

But the genuine difficulty has been used, systematically, as a reason not to try. And that is a different problem.

♦

The absence of the counterfactual has a specific structural consequence that is rarely stated plainly: it makes incompetence structurally invisible.

A physician who consistently makes poor decisions will see poor outcomes. But poor outcomes are common in medicine even when the care is excellent, because the underlying diseases are serious and the biology is complex. A physician with a higher-than-average complication rate can attribute the excess to patient complexity: sicker patients, more difficult cases, a referral pattern that concentrates the hardest problems in one practice. Without a rigorous case-mix adjustment across a large enough sample, distinguishing a physician whose outcomes are poor because of genuine patient complexity from one whose outcomes are poor because of substandard care is genuinely difficult.

The difficulty is real. It is also exploitable, because biology itself supplies plausible explanations for almost any clinical failure, and the system has no structural requirement to look behind a plausible explanation once it has landed. In a system where the profession evaluates its own performance, the difficulty of distinguishing genuine complexity from genuine incompetence provides structural protection for the incompetent. The noise provides cover, not just the biological noise of Chapter Four, but the professional noise of a system that has decided, structurally, that the difficulty of measurement is a sufficient reason not to measure.

The relationship between noise and anomaly is not symmetric. Noise hides the anomaly. But noise alone does not explain why the anomaly stays hidden once it is found. That requires structure, the system's decision, embedded in its architecture, about whether an anomaly that has been surfaced is treated as signal or returned to the background. Noise makes the anomaly hard to see. Structure decides whether seeing it is required.

The consequences of this structural invisibility are not abstract. In the early 2000s, a surgeon practicing at a regional Australian hospital had a prior licensing history in multiple American states. That history had included restrictions and complaints, none of which had generated a shared record, a cross-institutional signal, or a pattern visible to any system. Each jurisdiction had seen only its own slice. The outcomes that followed remained invisible to each institution he entered, because no institution maintained the comparative data that would have made them visible. The noise provided cover at every step: the deaths were attributed to surgical complexity, to the difficulty of the cases, to the inherent risk of operating on sick patients in under-resourced settings. Each attribution was individually plausible. Collectively, they concealed a pattern that had already been partially visible in another country.

The pattern became visible when a senior intensive care unit (ICU) nurse noticed it from her own ward, not because a system surfaced it, but because one person, watching one set of patients, recognized something that crossed the threshold of what complexity could explain. She raised her concerns internally. She was ignored. She was undermined. She eventually went outside the chain of command entirely, to an elected official, who raised the matter publicly. A formal public inquiry followed. What she did required extraordinary personal courage and came at significant professional cost to her. The structural lesson is not that she was brave, though she was. It is that the architecture required her courage to produce what it should have produced automatically: a pattern recognized, an anomaly treated as a verdict, a system forced to see what it could not see itself.

This is where the software analogy becomes most direct. In software, the argument that code review is too difficult because software is complex and requirements are ambiguous is not accepted as a reason not to review the code. The complexity is real. The difficulty is real. The review happens anyway, because the alternative — shipping code that has not been evaluated against expected behavior — is understood to produce outcomes worse than the difficulty of the review.

Medicine has largely made the opposite choice. The difficulty of evaluating physician performance against a counterfactual has been used as a reason to rely primarily on self-evaluation. The physician assesses the outcome. The physician determines whether the outcome was acceptable given the complexity of the case. The physician decides whether anything should change. The standard against which the performance is evaluated is, in most cases, the physician's own standard.

♦

Other performance-intensive fields have addressed this problem, imperfectly but deliberately, by separating the evaluation of performance from the performance itself.

Aviation tracks every serious incident involving every pilot across every airline. The data is anonymized to protect against punitive use, but it is analyzed systematically to identify patterns: a higher-than-average incidence of specific error types, a cluster of incidents under specific conditions, a pattern that suggests a systemic rather than individual problem. The goal is not to catch bad pilots. It is to find bad systems. And finding bad systems requires data that exists independently of the people whose systems are being evaluated.

Medicine has begun to move in this direction, slowly and unevenly. Hospital outcome databases. Physician profiling. Value-based payment models that tie reimbursement to measurable outcomes. These are genuine efforts to create the kind of comparative data that would make the counterfactual visible. They are also partial, inconsistent, and fiercely resisted.

The resistance comes in two forms. One is legitimate: the concern that comparative outcome data, without rigorous case-mix adjustment, will penalize physicians who treat the sickest patients and reward those who select the healthiest ones. This concern is real and the methodology for addressing it is genuinely difficult. It is also addressable, and the field of outcomes research exists precisely to address it.

The second form of resistance is less legitimate: the profession's long-standing claim that physician performance is too complex and too individualized to be captured by external metrics. This claim contains truth. It also contains the profession's interest in maintaining control over the criteria by which it evaluates itself. The two are not always easy to separate, which makes the claim very useful as a resistance strategy.

◆

Return to the question I asked in Chapter One. My symptom resolved. The file closed. Was I better?

I still do not know. And here is what I now understand more precisely about why I do not know: the system did not evaluate what happened to me against what should have happened. There is no record anywhere of what the expected outcome was for a patient with my presentation, treated with the sequence of interventions I received, at the institutions where I was treated, by the physicians who treated me. The outcome was observed. The counterfactual was not constructed. The gap between the two — the measure of whether the care I received was appropriate, adequate, or something else — was never calculated.

The calculation is not impossible. For many of the decisions made in my care, there is published evidence about expected outcomes under different treatment paths. The calculation would require someone to make it. No one in the system is currently required to make it. The physician's performance is evaluated against her own standard, by herself, without a systematic counterfactual. The outcome I experienced is absorbed into the aggregate of her practice, attributed to the complexity of my case, and filed without comparison.

In software, this would be called shipping unreviewed code. The code may be fine. It may not be. Without the review, no one knows. And in a field where the cost of not knowing falls on patients rather than on the physician, the structural incentive to not know is, all else being equal, stronger than the structural incentive to find out.

The counterfactual is absent. That is not a small thing. It is the mechanism by which every other high-stakes performance field holds its practitioners accountable. Its absence in medicine is not an accident of epistemology. It is a structural feature produced by decisions, made over decades, about who controls the criteria by which performance is evaluated, and whose interests those criteria have been designed to serve.

Without it, performance cannot be seen, only outcomes. And outcomes, in isolation, can always be explained.

Chapter Eight

The System That Cannot See Itself

Somewhere in a medical records system, there is a chart for my back problem. It records six visits. It records the imaging that was ordered and the findings that were reported. It records the physical therapy referrals and, presumably, some notation of the patient's progress or lack of it. And it records, in the final entry, something to the effect that the patient reported improvement and was instructed to return if symptoms recurred.

What the chart does not record is the question that was never asked. It does not record the diagnostic path that was not taken. It does not record the mechanism that was not investigated, the risk assessment that was not performed, the conversation about prognosis that did not happen. It does not record what was missed, because the chart is not a record of what happened. It is a record of what the system chose to document about what happened. Those are not the same thing.

What is not recorded cannot become evidence. What cannot become evidence cannot become error. The gap between those two sentences is where most of medicine's learning does not happen.

What cannot become error cannot become change.

If something had gone seriously wrong — if the symptom had returned as something worse, if a missed finding had later become significant — the chart would have been reviewed. The review would have found a sequence of reasonable decisions, each documented in standard language, each defensible against the standard of care. The review would not have found the question that was not asked, because the question was not asked, and the chart records questions only when they are asked.

This is the problem this chapter is about. Not the individual failure of my chart, which is ordinary and unexceptional. The structural feature of all charts, in all institutions, that makes the most important failures in medicine the hardest to see.

♦

In aviation, when something goes wrong, the investigators do not ask the crew what happened and stop there. They read the data. Crew interviews happen: witness accounts, statements, air traffic control recordings are all part of the investigation. But they are reconciled against an independent record the crew did not create. If the account diverges from the data, the data takes precedence.

The flight data recorder is a literal black box, fireproof, crash-resistant, designed to survive the conditions that destroy everything else on the aircraft. It captures hundreds of parameters in real time: airspeed, altitude, heading, engine thrust, control surface positions, vertical acceleration. The cockpit voice recorder captures the spoken words, the ambient sounds, the warnings and alarms of the final hours of the flight. Together they produce a record of what actually happened, not what the crew intended, not what the crew remembers, not what the crew would choose to emphasize in a post-incident report. What happened.

When investigators read the data, they are reading a record that was not created by the people being investigated. It was created by the aircraft itself, continuously, without judgment or selection, with no knowledge that something was about to go wrong and no interest in how the outcome would reflect on anyone. It is, as close as current technology permits, an objective record of the system's actual behavior.

In medicine, there is no black box. There is a chart. The chart is a record created by the people who provided the care, in real time, for a patient they were trying to help, without knowing how the case would end. It records what they noticed. It does not record what they missed. It records the decisions they made. It does not record the decisions they considered and rejected. It records the story they chose to tell about the encounter, a story shaped by professional norms, liability concerns, time pressure, and the entirely human tendency to frame events in ways that reflect reasonably on the person doing the framing.

The chart is also, inevitably, a legally consequential document. This is not incidental to how it is written: it is formative. A physician who records three competing diagnostic hypotheses has created a liability footprint. If she pursues hypothesis A and the patient suffers an adverse outcome because the reality was hypothesis C, a plaintiff's attorney will use that chart to prove negligence: you considered hypothesis C and did nothing to rule it out. The intellectual record of medicine is therefore shaped not just by what the system thinks to ask, but by what the system's legal architecture makes dangerous to write down. Narrative ambiguity is legally dangerous. The chart eliminates it systematically.

When something goes wrong and the chart is reviewed, it looks for departures from documented standards. It finds them, or it does not. What it almost never finds is the failure of recognition — the symptom attributed to the wrong cause, the test that was not ordered, the pattern that was present but not seen — because those failures leave no trace. The chart records what was thought. It does not record what should have been thought. The gap between the two is precisely where the most important failures live.

There is a failure mode that the chart is specifically designed to hide. A patient returns for the third time with the same complaint. Each previous visit is recorded: the symptom noted, the examination documented, the patient reassured. What the chart cannot record is the pattern. Not one visit, but three. Not one complaint, but the same complaint recurring in a system that has been unable to resolve it. The chart treats each encounter as an independent event. The patient's persistence is absorbed as evidence of her anxiety rather than as a signal requiring a different diagnostic approach. This is not always wrong: some patients do return with symptoms that have no organic cause, and reassurance is the appropriate response. The structural failure is not that the system reassures. It is that the system has no mechanism for distinguishing between the patient whose persistence is anxiety and the patient whose persistence is the only signal the system has not yet understood. Both receive the same note: patient reassured and counseled. The accumulation across encounters — the pattern that would reveal the difference — has no syntax in the chart. Each note closes its own encounter. The question the pattern demands is never asked.

The chart's incompleteness compounds at transitions. When a patient moves from an emergency department to an inpatient ward, from a hospital to a rehabilitation facility, from a specialist back to a primary care physician, the incoming provider inherits the chart. The chart conveys the documented story. It does not convey what was considered and rejected, what the previous provider suspected but could not prove, what the patient said that was not thought significant enough to record. Each hand-off is a reset of the clinical narrative to what the chart contains, and in a system where the chart systematically omits the questions that were not asked, each transition multiplies the omission. The patient who has been seen by four providers carries a chart that four providers helped create. None of them wrote down what they missed. The accumulation of their silences is invisible to the fifth. And if that patient's problem crosses specialty boundaries — cardiology and nephrology and primary care all seeing their own piece — the accumulation of their separate silences is not visible to any of them.

♦

The consequence is that medicine's errors are systematically underrepresented in its own records. Not by a small margin.

A landmark study released in 1999 by the Institute of Medicine (IOM), titled To Err is Human, estimated that between 44,000 and 98,000 Americans died each year from preventable medical errors. Later estimates, most prominently a 2016 analysis in the British Medical Journal, have placed the figure above 250,000 deaths per year, a claim that, if accepted, would make medical error one of the leading causes of death in the United States.

These estimates are contested, and the contest is legitimate. Counting medical errors requires judgments about what was preventable, what constitutes an error rather than an acceptable complication, and how to attribute a death in a patient with multiple serious conditions. The numbers are not precise.

But the uncertainty cuts in one direction. The chart-based reconstruction that underlies most estimates of medical error systematically undercounts errors of omission: the test that was not ordered, the diagnosis that was not considered, the pattern that was not recognized. These failures do not appear in the chart because they were never documented. A review of the chart cannot find what the chart does not contain. The estimates of preventable deaths are therefore almost certainly undercounts. The true number is not known, and the system is not currently designed to find out.

♦

Medicine's inability to see its own errors is not a problem of insufficient data. It is a problem of architecture.

A system that generates its own safety data, through its own processes, evaluated by its own members, will generate data that reflects the system's understanding of itself. Not an independent assessment. Not an external measurement. A reflection. And a reflection, by definition, shows you only what you choose to show yourself.

The aviation model is different in one structural feature that matters more than any other: the people who investigate aviation accidents are not the people who fly the planes. The National Transportation Safety Board employs investigators whose sole function is to understand what happened and to recommend changes. They are not part of the airline. They are not part of the manufacturer. They have no relationship to the crew and no financial interest in the outcome. When they read the black box data, they are reading a record created by the aircraft, not by the pilots, and they are reading it as outsiders whose institutional interest is finding the failure, not minimizing it.

There is one more distinction worth making precise. We cannot have a literal flight data recorder for a human body. The biology is too opaque and the variables too vast. No two patients are built identically, and the failure modes were not documented by an engineer. But we do not need to audit the biology. We need to audit the decisions. A rigorous independent investigation of a medical adverse event does not ask why the body failed. It asks whether the clinical team's decision-making process was rational given what was known, whether the filters allowed the right information to reach the right people at the right time, and whether the diagnostic paths that were rejected were rejected for documented reasons. We cannot record every biological fluctuation. We can require a record of why a specific hypothesis was considered and why it was set aside. We cannot audit the body. We can audit the mind that was responsible for understanding it.

Medicine has produced its own oversight in the form of accreditation bodies, peer review organizations, and professional licensing boards. These bodies do genuine work. The Joint Commission has improved safety practices in measurable ways. Peer review has caught some errors and disciplined some practitioners. The licensing boards have removed some physicians who should not be practicing. None of this is nothing.

But none of it is independent in the way that matters. The bodies that oversee medicine are staffed by people who came from medicine, shaped by standards that medicine set, evaluating performance against criteria that medicine defined. They are not external. They are internal, in a different form. They are the profession regulating itself through a slightly more formal version of self-regulation. And the structural consequence is the same: the system regulates to the standard it is comfortable with, not necessarily to the standard that would surface the errors it most needs to see.

The system cannot see itself because it is looking in a mirror it made. The mirror shows what the system chose to put in front of it. The failures that would require looking somewhere else — at independent data, at external investigators, at records not created by the people being evaluated — remain in the dark.

♦

What would it take for the system to see itself?

The minimum requirement is a record that is not created by the people being evaluated. Not a chart that documents the story the provider chose to tell, but a parallel record — of outcomes, of complications, of deviations from expected trajectories — created independently and compared systematically to what the chart says happened.

The second requirement is an investigation process that is genuinely external. Not a mortality and morbidity conference run by the hospital whose practices are being reviewed. Not a peer review conducted by colleagues who share the same training, the same norms, and the same professional interests. An investigation conducted by people whose institutional interest is finding the failure, not explaining it away.

These requirements are not new. They have been articulated by patient safety researchers for decades. The gap between articulation and implementation is not epistemological. The knowledge of what to do has been available for a long time. The gap is structural, produced and maintained by the same forces that keep the chart incomplete, the filter asymmetric, and the counterfactual absent.

My chart records that I reported feeling better and was told to return if symptoms recurred. It does not record whether the care I received was appropriate, whether the diagnostic process was adequate, or whether the outcome I experienced reflects what should have happened or what happened by default. The system cannot answer those questions because it is not designed to ask them. It is designed to record the story. The story is not the same as the truth.

Chapter Nine

Designing for the Anomaly

The previous chapters described what the system is. This one asks what it could be. Not as an ideal, but as a practical question with concrete answers, answers that, in specific domains and specific institutions, have already been demonstrated to work.

The demonstration matters. One of the defenses of the current structure is that the problems described in this book are intractable: that the biology is too complex, the variation too large, the system too entrenched to be meaningfully redesigned. That defense is not supported by the evidence. When medicine has built structures that surface anomalies, those structures have worked. The question is not whether redesign is possible. It is why the redesign has been so partial, so slow, and so fiercely resisted.

♦

In the early 2000s, a physician named Peter Pronovost at Johns Hopkins identified a preventable pattern: patients in intensive care units were dying of infections caused by bacteria introduced during the insertion of central venous catheters. Not because the physicians were incompetent. Not because the hospitals were negligent. Because the process for inserting catheters was not standardized, and in the absence of standardization, steps were missed, surfaces were contaminated, and infections resulted.

Pronovost developed a checklist, what came to be known as the central line bundle. Five items. Wash hands. Clean the patient's skin with chlorhexidine. Use full-barrier precautions. Avoid the femoral site if possible. Remove unnecessary catheters promptly. None of these steps was unknown. All of them were taught in medical training. In practice, under the pressure and distraction of an ICU, one or more of them was routinely missed.

When the checklist was implemented rigorously in Michigan ICUs — not just distributed, but enforced, with nurses empowered to stop a procedure if a step was skipped — catheter-related bloodstream infections fell sharply within three months and were down by sixty-six percent over eighteen months. Over that period, the intervention was associated with an estimated 1,500 lives saved in Michigan alone. The anomaly had been made visible. Made visible, it could be eliminated.

Nothing new had been discovered. Only something old had been made impossible to ignore.

That is what design does. It changes what the system is allowed to ignore.

This is what designing for the anomaly looks like in practice. Not a technological breakthrough. Not a new drug or a new procedure. A structural change: the decision to treat a class of adverse events as anomalies the system was obligated to prevent, rather than as complications it was permitted to accept.

♦

The same logic produced similar results in the operating room. Atul Gawande and colleagues, working with the World Health Organization, developed a nineteen-item surgical safety checklist, a structured pause before the incision to confirm that the right patient was on the table, the right site was marked, the necessary equipment was available, and the team had communicated about anticipated risks. The checklist was piloted in eight hospitals across eight countries, spanning rich and poor, urban and rural.

Complications fell by thirty-six percent. Deaths fell by forty-seven percent. The effect was not confined to low-resource settings. It was consistent across all eight sites. Here, again, the mechanism was not a new technology. It was the structural decision to treat the preventable surgical complication as an anomaly rather than an inevitability, to ask, before every operation, whether the conditions were in place for the procedure to go as planned.

Both interventions (the central line bundle and the surgical checklist) share a structural feature that is more important than their specific content. They were designed to surface the anomaly before it occurred. Not to review it afterward. Not to attribute it to patient complexity once it had happened. To identify, in real time, the conditions that made the anomaly likely, and to interrupt those conditions before they produced harm. This is the design principle. It is simple. It is generalizable. It is not how most of medicine is currently organized.

♦

There is one domain within medicine where the transformation was not partial and not local. Anesthesia mortality in the United States in the early 1980s — variously estimated at 1 to 2 per 10,000 anesthetics — represented thousands of preventable deaths and cases of serious brain damage each year. Today it is approximately 1 in 200,000, a more than twentyfold reduction that the Institute of Medicine, in its landmark 1999 patient safety report, cited as the field's most impressive safety achievement. The question worth asking forensically is not what changed, but what forced it to change, because the answer to that question is precisely the framework this book has been building.

♦

The forcing anomaly arrived from outside the profession. In 1982, ABC News broadcast a documentary called "The Deep Sleep," which detailed anesthesia deaths in terms that made them undeniable to a general audience. Malpractice premiums for anesthesiologists had already reached crisis levels: the specialty constituted three percent of physicians but generated twelve percent of malpractice payouts. The anomaly was visible, persistent, and financially undeniable. What the broadcast did was make it publicly undeniable as well. The profession could no longer absorb it. Dr. Ellison Pierce, then president of the American Society of Anesthesiologists (ASA), used that pressure to do something that distinguished this case from nearly every other in medicine: he converted the individual pressure into structural change. He created the ASA Committee on Patient Safety, which mandated monitoring standards — pulse oximetry, capnography, continuous presence of a qualified anesthesiologist — across the entire specialty. The standards were adopted nationally in 1986. Malpractice premiums fell sixty-six percent over the following five years. Mortality fell twentyfold. The profession received those numbers as proof that the architecture had worked.

♦

Why did anesthesia succeed where most of medicine has not? Three conditions aligned that rarely align elsewhere. The anomaly was externally forced into visibility: not by a research paper in a specialty journal, but by a television broadcast to a lay audience and a malpractice crisis that made the financial cost of inaction greater than the cost of change. The mechanism was identifiable: deaths were caused by late recognition of specific, preventable events — esophageal intubation, hypoxia, ventilator failure — that monitoring technology could catch in real time. And the response was made structural rather than personal: not a brave individual insisting on handwashing in one ward, but a specialty-wide standard that required the monitoring regardless of who was in the room or what the local institution preferred. Semmelweis was right but his intervention was personal and fragile. Pierce was right and his intervention was institutional and durable. The difference was not courage. It was architecture.

♦

These successes are genuine. They are also circumscribed. They work in settings where the link between a specific process and a specific outcome is tight and measurable, where the anomaly is defined narrowly enough that its conditions can be identified and standardized. Central line infections. Surgical complications from missed steps. Wrong-site surgery. These are categories of harm where the mechanism is known, the intervention is specifiable, and the outcome is attributable.

The broader structural failures of medicine — the ones the preceding eight chapters have described — do not fit this template. The question not asked at the end of my back pain encounter does not have a checklist solution. The evidence deferred for forty-two years while sailors died of scurvy was not a missed step in a standard process. The filter that passes favorable findings and holds unfavorable ones is not addressable by a structured pause before a procedure. The structural failures are systemic. The checklist interventions are local. Both are real. They are not the same problem.

♦

What would a system-level structural commitment look like?

The first requirement is independent investigation of serious adverse events. Not mortality and morbidity conferences run by the institutions whose practices are being reviewed. Not peer review conducted by colleagues who share the same training and professional interests. An independent body — structured like the National Transportation Safety Board, functionally separate from the profession and the industry — whose explicit mandate is to investigate serious failures and to publish findings that the system is required to act on. The investigation would not be designed to assign blame to individuals. It would be designed to identify system failures. Blame-based investigation produces concealment. System-based investigation produces information.

Aviation built this system after its own era of unacceptable loss. Aviation took thirty years to reduce its version of that number by more than ninety-nine percent, not through better pilots or better aircraft alone, but through a structural decision to treat every serious incident as a system failure. And aviation had a structural reason to make that decision. When an airliner crashes, the cost is immediate and concentrated on the operators of the system: a destroyed asset, an instant market reaction, regulatory scrutiny, careers ended. The operators bleed when the system fails. Medicine has no equivalent. When a patient dies of preventable error, the operators do not bleed; the patient does. The NTSB was empowered to investigate independently. The flight data recorder was mandated. Near-misses were reported to a confidential federal database, analyzed for patterns, and translated into required changes. By widely cited estimates, medicine contributes to many tens of thousands — possibly more than 250,000 — American deaths from preventable errors each year, though the figure remains contested. The system still lacks a stable, universally accepted way to count preventable harm. The difference is not complexity. The difference is structural.

The second requirement is comparative outcome data that is genuinely comparable. Not the fragments that currently exist: hospital report cards that are contested on methodology, physician profiles that lack adequate case-mix adjustment, outcome databases that cover some procedures and not others. A systematic infrastructure of outcome measurement, produced by independent analysts, adjusted rigorously for patient complexity, and published in a form that patients, referring physicians, and policymakers can actually use.

The third requirement is a legal structure that does not punish disclosure. The current structure creates strong incentives for concealment: an institution that acknowledges an error increases its legal liability. The institution that conceals it, and attributes the outcome to patient complexity, faces less exposure. This incentive is directly opposed to the learning that produces safety. Several states and countries have experimented with communication-and-resolution programs that decouple acknowledgment from liability, allowing institutions to be honest about errors without automatically triggering malpractice claims. The evidence from these programs is consistent: they reduce litigation, improve patient satisfaction, and produce the honest error analysis that generates systemic learning. They have not been widely adopted, because their adoption requires the legal and insurance industries to accept changes that reduce their revenue from adversarial litigation.

The evidence that this works exists. The University of Michigan Health System implemented exactly this model in the early 2000s: open disclosure of adverse events, independent investigation of each case, and proactive compensation when care fell below standard. Over the following years, lawsuits fell by approximately two-thirds, new claims by more than a third, and total liability costs by more than half; legal defense costs dropped substantially and the time required to resolve claims shortened significantly. Patient safety event reporting increased by nearly forty percent as staff became willing to report errors rather than conceal them. The Michigan Model has since been replicated at multiple institutions with consistent results. What it demonstrates is not that disclosure is risk-free — some cases do proceed to litigation — but that the incentive structure that currently rewards concealment is not an immovable feature of the legal landscape. It is a choice. Other choices produce different outcomes. The Michigan Model has been available as evidence for more than two decades. It has not been widely adopted. The resistance is not epistemological.

♦

None of these changes requires knowledge that does not exist. The interventions have been piloted. The evidence is in the literature. What is missing is not the design: it is the structural commitment to implement it against the interests that benefit from the current arrangement.

The resistance is real and specific. It comes from the profession, which loses autonomy. From the institutions, which lose control of their own narrative. From the legal industry, which loses the revenue model of adversarial litigation. From the political system, which has repeatedly declined to impose on medicine the kind of external accountability it imposes on other industries whose failures kill people.

The ICUs where the infection rates went to zero are proof that the design is possible. When the leadership changes, when the incentives shift, when the pressure subsides, the rates can return. What the checklist alone cannot produce is the architecture that makes the anomaly impossible to ignore regardless of who is in charge.

That architecture is what the evidence demonstrates is possible. Not better individuals. Not better intentions. A system designed so that the anomaly – the question not asked, the evidence deferred, the signal hidden, the error unrecorded – cannot stay hidden. Not because someone was courageous enough to surface it. Because the structure required it to be surfaced.

Chapter Ten

Trust, With Eyes Open

I trusted the system. I want to be clear about that, because the argument of this book could be read as an argument for distrust, and it is not.

I went to six visits. I followed the referrals. I did the physical therapy. I got the imaging when it was ordered. When the specialist required a new X-ray before she would see me, I argued with my wife, lost the argument on rational grounds, and got the X-ray. When the sixth visit ended with the instruction to come back if symptoms returned, I accepted the instruction and left. I trusted that the system was doing what it was designed to do, that the people within it were competent and well-intentioned, and that the process I had moved through was approximately what the evidence supported.

Much of that trust was justified. The physicians who treated me were not negligent. The protocols they followed were not irrational. The imaging they ordered was consistent with standard practice. Nothing in my case suggests that I received care that was obviously wrong.

What I received was care that was structurally incomplete. The question was not asked. The counterfactual was not constructed. The record did not capture what was missed. The outcome was attributed to the patient's self-report and the file was closed. None of this required incompetence or bad faith. It required only that the system work exactly as it was designed to work. And I trusted that design without fully understanding what it was.

That is the problem this chapter is about. Not the absence of trust, but the character of it.

♦

The argument of this book is not that medicine does not work. It works. It works remarkably, in many domains, for many patients, across an enormous range of conditions. The eradication of smallpox.

The near-elimination of polio. The development of antibiotics that transformed bacterial infections from death sentences into inconveniences. The transformation of HIV from a uniformly fatal disease into a manageable chronic condition. The survival rates for childhood cancers that would have been unthinkable fifty years ago. These are real, irreversible goods. They are the products of genuine scientific progress, genuine clinical skill, and genuine institutional investment over more than a century.

♦

Trust is not the same as faith.

Faith asks for belief in the absence of evidence, or, more precisely, independent of the evidence. It is a commitment that does not update in response to what it finds. Faith in medicine would mean believing that the care you receive is appropriate regardless of whether the system has mechanisms for knowing whether it is appropriate. It would mean trusting the physician's confidence regardless of whether the confidence is warranted by the evidence.

Trust is different. Trust is confidence grounded in evidence, including evidence about how the trusted party behaves when things go wrong. A person you trust to handle money is someone you have reason to believe will handle it honestly, not merely someone who assures you they will. The assurance is not the evidence. The track record, the transparency, the accountability structures that would surface dishonesty if it occurred: those are the evidence.

It is also the difference between credentialing — the front-loaded trust structures of admission, certification, and institutional pedigree — and continuous verification, the architecture through which performance is repeatedly tested against external reality. Other high-risk systems evolved toward the second. Medicine remains unusually dependent on the first.

It is not given. It is built from what the system does when it fails.

A medicine designed to surface its errors, to investigate them independently, to learn from them systematically, and to hold itself publicly accountable for that learning would deserve a high level of trust. Not unconditional trust — no institution deserves that — but the kind of confident trust that a patient is currently asked to extend without the evidence that would warrant it. A medicine that is designed, structurally, to absorb its errors, defer inconvenient evidence, and resist external accountability deserves a more qualified trust. Not distrust. Trust with open eyes.

♦

What does trust with open eyes look like in practice?

It looks like a patient who, before a procedure, can ask what the complication rate is at this institution, with this surgeon, for patients with this presentation, and receives a meaningful answer rather than a reassurance. Not "we have excellent outcomes" but "our thirty-day complication rate for this procedure is X, which compares to the national average of Y." The difference between those two responses is the difference between a system that trusts its patients with the evidence and a system that manages them with confidence.

It looks like a physician who says "I don't know" when she does not know. The physician who constructs a confident explanation for every outcome — who attributes every unexpected result to the complexity of the disease, who never acknowledges uncertainty, who frames every outcome as consistent with what was expected — is not practicing with greater competence than the physician who acknowledges the limits of what is known. She is performing competence rather than demonstrating it. And performances, over time, erode the trust they are designed to maintain. The patient who is told confidently that the symptom is nothing, and then finds it is something, does not conclude that medicine is infallible and this was an exception. She concludes that confidence is a style, not a signal.

It looks like a system that treats the patient's experience as evidence rather than feedback. Not "please rate your satisfaction with your care" but "your symptom resolved — what was the mechanism, and what does the resolution tell us about the underlying condition?" The first is a customer service question. The second is a clinical one. The gap between them is the gap between a system oriented toward patient satisfaction and a system oriented toward patient health.

♦

The medicine that exists today asks patients for a kind of trust that the medicine that exists today has not yet fully earned.

It asks for confidence in a system that does not systematically measure whether the care it provides matches what the evidence supports. It asks patients to move through a system of filters and processes without access to the comparative data that would allow them to evaluate whether what they are receiving is appropriate. It asks, in short, for faith, and offers, in return, the assurance of the people providing the care.

But the patient who trusts that architecture without understanding its limitations is not exercising judgment. She is delegating it. A patient with no access to comparative data, no independent verification of performance, no mechanism for evaluating whether the care she is receiving is appropriate, is not a participant in her own health. She is a subject of a system she cannot see.

The patients who force the system to see what it missed — who persist past the point where the file would be closed, who bring the pattern to a clinician willing to look — are doing, as patients, what the architecture should do automatically. Their persistence is not a feature of the system. It is a workaround for the absence of one.

♦

This book is not asking for distrust.

The medicine that would deserve the trust patients currently extend to it would be one that had made its errors visible, investigated them honestly, learned from them systematically, and changed its behavior in response. That medicine does not yet exist in full. It exists in fragments: in the ICUs where the checklist drove infections to near zero, in the institutions where honest mortality review changed practice, in the physicians who say what they do not know and mean it. These are evidence that such medicine is possible.

I went to six visits. I paid fifteen dollars each time. I trusted the system to ask the questions it knew how to ask. It did. I trusted it also to ask the questions it did not know how to ask, or was not designed to ask, or had no incentive to ask. It did not. The difference between those two forms of trust is the difference this book has spent ten chapters trying to describe.

Chapter Eleven

The Anomalies Around Us

There is a list, maintained by the Choosing Wisely campaign since 2012, of medical practices that the evidence suggests provide little or no benefit to patients. The list has been compiled by physician specialty societies across dozens of fields. It runs to hundreds of items. It is the profession's own accounting of its own practices.

The difficulty is no longer seeing it. The difficulty is being required to act on it.

The anomalies around us now are less visible, not because they are smaller, but because they are diffuse. They do not cluster around a single ward or a single bacterium. They are embedded in the ordinary practice of medicine, distributed across millions of encounters, absorbed into the background noise of a system whose errors are not counted and whose costs are not attributed.

What makes them anomalies in this book's specific sense is not that they are unknown. Each of them has been documented, named, and published in peer-reviewed literature. Each has been present for decades. Each has been acknowledged by the profession that produces it. They meet the three conditions this book has been describing: they are visible, they are persistent, and they have not been treated as verdicts. The system has developed, for each of them, a characteristic response, a way of absorbing the anomaly that falls short of requiring structural change. That response is not ignorance. It is architecture.

Low-value care is the first. Arthroscopic knee surgery for osteoarthritis has been shown in randomized trials to be no more effective than sham surgery. Antibiotics for uncomplicated viral respiratory infections do not treat viral infections and contribute to antibiotic resistance. The scale is not trivial: estimates of the annual cost of low-value care in the United States range from $75 billion to $100 billion.

The second anomaly is overdiagnosis: the detection of disease that would never have caused symptoms or death if left undetected. What is absent is the clinical significance. The patient whose small papillary thyroid cancer is discovered incidentally on an ultrasound ordered for a different reason has a real finding. She also has, with very high probability, a finding that will never harm her.

Thyroid cancer diagnosis in the United States increased dramatically over three decades following the widespread adoption of cervical ultrasound, driven almost entirely by detection of small papillary tumors that would previously have gone unnoticed. Mortality from thyroid cancer did not change. The enthusiasm for detecting disease has not been matched by an equal enthusiasm for asking what detection costs the patients who are detected.

The third anomaly is the most structurally revealing, because it does not require asking whether a specific treatment is harmful. It requires only asking why two patients with identical clinical presentations receive dramatically different care depending on where they live.

For most conditions in medicine, there is substantial geographic variation in the rates at which specific procedures are performed, variation that persists after adjusting for differences in patient age, severity of illness, and other clinical characteristics. A patient with lower back pain in one region of the United States is roughly twice as likely to have surgery as a patient with an identical presentation in another region. A patient with early-stage prostate cancer has a significantly different probability of surgery depending on which urologist he sees, independent of the clinical features of the cancer. This variation has been documented continuously since the 1970s by the Dartmouth Atlas of Health Care, which has tracked regional differences in Medicare spending and procedure rates across the United States for nearly half a century. The documentation is unambiguous, the methodology is rigorous, and the findings have been replicated across conditions, regions, and time periods. The Dartmouth Atlas is not a study. It is an archive of a persistent anomaly that the system has had forty years to convert into a verdict and has not. The variation persists not because the evidence is uncertain but because the system contains no mechanism that requires the variation to be explained. In aviation or civil engineering, unexplained variation in failure rates across comparable facilities would constitute a system failure requiring mandatory investigation. In medicine, it has constituted, for forty years, a research finding.

The variation cannot be fully explained by patient preferences, clinical uncertainty, or regional differences in disease burden. Much of it reflects local practice norms, habits transmitted through training and reinforced by the absence of any mechanism requiring them to be evaluated against outcomes. The implication is simple: at least one of those treatments is unnecessary. The system does not require that conclusion to be drawn.

There is a fourth anomaly, larger than the three described above, that subsumes them all. The United States spends more on healthcare per capita than any other country in the world, nearly twice the median of comparable developed nations. Its outcomes, measured by life expectancy, infant mortality, and the prevalence of treatable chronic conditions, rank below most of those nations. The gap between spending and outcomes is not small, not recent, and not explainable by population heterogeneity or lifestyle factors alone. It is the closest thing the system has to a system-level eight arcminutes: a number that does not fit, that has persisted for decades, and that the current framework has explained around rather than resolved. It is the forcing anomaly the system has not yet been required to treat as a verdict.

Four anomalies, visible in the current system, documented by the profession's own research apparatus, persistent across decades of measurement. None has become a verdict. Each has instead generated a characteristic institutional response: a campaign, a report, a guideline, an atlas, a commission. Each response falls short of the structural requirement that would make the anomaly impossible to continue. The Choosing Wisely list names the problem. It does not require the problem to stop. The Dartmouth Atlas documents the variation. It does not require the variation to be explained. The IOM report quantified preventable harm. It did not create the mechanism that would prevent it. In each case, the system has found a way to see the anomaly that stops short of being required to act on it. That is not failure of knowledge. It is the structure of the system expressing itself. The system also cures diseases that killed our grandparents, extends lives that previous generations could not save, and produces genuine miracles at the individual level every day. None of that exempts it from this accounting. A system can be admirable in what it does and still be structured so that what it does not do is invisible to it. That is not a contradiction. It is precisely the structural condition this book has been describing.

The historical anomalies — Semmelweis, scurvy, Marshall — show the same structure. The structure that failed those historical patients is still present. The patients absorbing the cost of the current anomalies are not historical. They are alive now, moving through a system that can see its own failures and has chosen, structurally, not to treat them as verdicts.

Conclusion

The Error That Must Be Seen

There is a calendar maker in this story, though he does not appear until the end.

In 1825, Lieutenant-Colonel John Warren of the British East India Company encountered a calendar maker in Pondicherry who calculated the timing of a lunar eclipse without instruments, without books, and without written tables. He arranged shells on the ground and computed from mathematical relationships memorized in verse. His prediction was off by four minutes. Warren was astonished. The tradition that produced this calendar maker had been unbroken for thirteen centuries.

The first part of this book uses this story to illustrate what a working knowledge system looks like at its best: precise, transmissible, self-sustaining, producing results across an extraordinary span of time. The calendar maker is not a failure. He is an achievement. He is also, in that context, standing at the edge of a limit. The tradition that produced him was not designed to see the questions it had not yet asked.

Medicine is standing at a similar edge. Not the same edge: medicine is not the Kerala School, and the analogy should not be pressed too far. But the structural question is the same. A system can be excellent at what it is designed to do and still have structural features that prevent it from seeing its own limits. The calendar maker computed precisely. The tradition that formed him did not produce the heliocentric model. Medicine cures diseases that killed our grandparents. The structure that produced those cures is not designed to surface the errors that persist within it.

What this book argues is narrower than “medicine should change.” It argues that the conditions for mandatory learning are absent in a domain where the cost of their absence is paid by patients who did not choose it. That is the structural fact. What follows from it is for others to decide.

♦

The Institute of Medicine named it in 1999. Peter Pronovost and his colleagues demonstrated, in the ICUs of Michigan, that when the system makes a structural commitment to treating a class of adverse events as anomalies it must prevent, the anomalies fall.

♦

Kepler asked the same question. He refused to treat eight arcminutes as a rounding error. He insisted, against the weight of thirteen centuries of precision, that the anomaly was a verdict rather than a nuisance. The cosmos broke. The new model followed.

Medicine has its Keplers. Semmelweis insisted. Marshall insisted. Pronovost built a checklist and refused to accept that catheter infections were an inevitable cost of intensive care. Each of them treated the anomaly as a verdict when the system was designed to absorb it. Each of them was right. Consider what happened when Semmelweis left Vienna: with the handwashing requirement no longer enforced, mortality in the First Division rose again. The individuals who had maintained the handwashing requirement were gone. The structure that had never required it reasserted itself immediately. The problem had never been the physicians. It had always been the architecture. And none of them — not even the ones who eventually prevailed — changed the architecture that made their insistence necessary in the first place.

♦

I went to six visits. I paid fifteen dollars each. The question was not asked. The file closed.

I do not know whether I was well served. I have no way to know, because the system did not construct the counterfactual that would allow me to know. I do not know whether the X-ray ordered at the first visit was clinically indicated for my presentation.

What I know is that the system processed me. It generated six encounters, each individually defensible, none of them connected by the question that a scientist would have asked: what is the mechanism, what does the resolution tell us, what is the risk going forward, and what would we expect to see if the underlying cause persisted? Those questions are not on any checklist. They are not reimbursable. They do not appear in the guidelines. They exist in the space the system has not yet decided to inhabit.

The patients who need that question answered are not waiting in the past.

They are inside the system while it continues.

The question is no longer whether the anomaly exists. It is whether the system will ever be required to see it. Until it is required, it will not.

Sources

What follows is a starting point rather than a comprehensive bibliography. The book draws on a wider literature than can be cited here; these are the works that most directly anchor the historical claims, technical material, and translated passages. Readers interested in pursuing any thread further will find these the natural points of entry.

Indian Mathematics and Astronomy

Colebrooke, Henry Thomas. *Algebra, with Arithmetic and Mensuration, from the Sanscrit of Brahmegupta and Bhāscara.* London: John Murray, 1817. Reprinted Delhi: Sharada Publishing House, 2008.

Doniger O'Flaherty, Wendy, trans. *The Rig Veda: An Anthology: One Hundred and Eight Hymns, Selected, Translated and Annotated.* Penguin Classics. Harmondsworth: Penguin Books, 1981.

Joseph, George Gheverghese. *The Crest of the Peacock: Non-European Roots of Mathematics.* 3rd ed. Princeton: Princeton University Press, 2011.

Patwardhan, K. S., S. A. Naimpally, and S. L. Singh. *Līlāvatī of Bhāskarācārya: A Treatise of Mathematics of Vedic Tradition.* Delhi: Motilal Banarsidass, 2001.

Pingree, David. *Jyotiḥśāstra: Astral and Mathematical Literature.* A History of Indian Literature, vol. VI, fasc. 4. Wiesbaden: Otto Harrassowitz, 1981.

Plofker, Kim. *Mathematics in India.* Princeton: Princeton University Press, 2009.

Rajagopal, C. T., and M. S. Rangachari. "On an Untapped Source of Medieval Keralese Mathematics." *Archive for History of Exact Sciences* 18, no. 2 (1978): 89–102.

Ramasubramanian, K., and M. S. Sriram. *Tantrasaṅgraha of Nīlakaṇṭha Somayājī.* Sources and Studies in the History of Mathematics and Physical Sciences. New Delhi: Hindustan Book Agency, 2011; London: Springer, 2011.

Sarma, K. V. *A History of the Kerala School of Hindu Astronomy.* Hoshiarpur: Vishveshvaranand Institute, 1972.

Sarma, K. V., ed. and trans., with explanatory notes by K. Ramasubramanian, M. D. Srinivas, and M. S. Sriram. *Gaṇita-Yukti-Bhāṣā (Rationales in Mathematical Astronomy) of Jyeṣṭhadeva*. 2 vols. New Delhi: Hindustan Book Agency, 2008; London: Springer, 2009.

Shukla, K. S., and K. V. Sarma, eds. and trans. *Āryabhaṭīya of Āryabhaṭa, with the Commentary of Bhāskara I*. New Delhi: Indian National Science Academy, 1976.

European Astronomy and the Scientific Revolution

Caspar, Max. *Kepler*. Translated by C. Doris Hellman. New York: Dover, 1993. (Originally published 1948.)

Christianson, John Robert. *On Tycho's Island: Tycho Brahe, Science, and Culture in the Sixteenth Century*. Cambridge: Cambridge University Press, 2000.

Gingerich, Owen. *The Eye of Heaven: Ptolemy, Copernicus, Kepler*. New York: American Institute of Physics, 1993.

Koestler, Arthur. *The Sleepwalkers: A History of Man's Changing Vision of the Universe*. London: Hutchinson, 1959.

Stephenson, Bruce. *Kepler's Physical Astronomy*. Princeton: Princeton University Press, 1994.

Swerdlow, Noel M., and Otto Neugebauer. *Mathematical Astronomy in Copernicus's De Revolutionibus*. 2 vols. New York: Springer-Verlag, 1984.

Voelkel, James R. *The Composition of Kepler's Astronomia Nova*. Princeton: Princeton University Press, 2001.

History and Philosophy of Science

Kuhn, Thomas S. *The Structure of Scientific Revolutions*. Chicago: University of Chicago Press, 1962. 4th ed., with introductory essay by Ian Hacking, 2012.

Lakatos, Imre. "Falsification and the Methodology of Scientific Research Programmes." In *Criticism and the Growth of Knowledge*, edited by Imre Lakatos and Alan Musgrave, 91–196. Cambridge: Cambridge University Press, 1970.

Narasimha, Roddam. "Axiomatism and Computational Positivism: Two Mathematical Cultures in Pursuit of Exact Sciences." *Interdisciplinary Science Reviews* 28, no. 1 (2003): 54–66.

Taleb, Nassim Nicholas. *The Black Swan: The Impact of the Highly Improbable.* New York: Random House, 2007.

Medicine and Medical Error

Blumenthal, David, et al. *Mirror, Mirror 2024: A Portrait of the Failing U.S. Health System — Comparing Performance in 10 Nations.* New York: The Commonwealth Fund, 2024.

Gawande, Atul. *The Checklist Manifesto: How to Get Things Right.* New York: Metropolitan Books, 2009.

Haynes, Alex B., Thomas G. Weiser, William R. Berry, et al. "A Surgical Safety Checklist to Reduce Morbidity and Mortality in a Global Population." *New England Journal of Medicine* 360, no. 5 (2009): 491–499.

Institute of Medicine, Committee on Quality of Health Care in America. *To Err Is Human: Building a Safer Health System.* Edited by Linda T. Kohn, Janet M. Corrigan, and Molla S. Donaldson. Washington, DC: National Academy Press, 2000.

Institute of Medicine, Committee on Quality of Health Care in America. *Crossing the Quality Chasm: A New Health System for the 21st Century.* Washington, DC: National Academy Press, 2001.

Kachalia, Allen, Samuel R. Kaufman, Richard Boothman, et al. "Liability Claims and Costs Before and After Implementation of a Medical Error Disclosure Program." *Annals of Internal Medicine* 153, no. 4 (2010): 213–221.

Lagasse, Robert S. "Anesthesia Safety: Model or Myth? A Review of the Published Literature and Analysis of Current Original Data." *Anesthesiology* 97, no. 6 (2002): 1609–1617.

Lind, James. *A Treatise of the Scurvy.* Edinburgh: Sands, Murray, and Cochran, 1753. Bicentenary reprint, edited by C. P. Stewart and Douglas Guthrie. Edinburgh: Edinburgh University Press, 1953.

Makary, Martin A., and Michael Daniel. "Medical Error — The Third Leading Cause of Death in the US." *BMJ* 353 (2016): i2139. (The "third leading cause" framing has been disputed in the subsequent literature.)

Marshall, Barry J., and J. Robin Warren. "Unidentified Curved Bacilli in the Stomach of Patients with Gastritis and Peptic Ulceration." *The Lancet* 323, no. 8390 (1984): 1311–1315.

Nuland, Sherwin B. *The Doctors' Plague: Germs, Childbed Fever, and the Strange Story of Ignác Semmelweis.* Great Discoveries series. New York: W. W. Norton, 2003.

Pronovost, Peter, Dale Needham, Sean Berenholtz, et al. "An Intervention to Decrease Catheter-Related Bloodstream Infections in the ICU." *New England Journal of Medicine* 355, no. 26 (2006): 2725–2732.

Shrank, William H., Teresa L. Rogstad, and Natasha Parekh. "Waste in the US Health Care System: Estimated Costs and Potential for Savings." *JAMA* 322, no. 15 (2019): 1501–1509.

Topol, Eric J. "Failing the Public Health — Rofecoxib, Merck, and the FDA." *New England Journal of Medicine* 351, no. 17 (2004): 1707–1709.

Wennberg, John E. *Tracking Medicine: A Researcher's Quest to Understand Health Care.* New York: Oxford University Press, 2010.

About the Author

Phani Kurada holds a PhD in molecular biology. He has spent three decades working across scientific research and software systems. His sustained interest in the epistemology of traditions and the history of Indian mathematical astronomy finds its first book-length expression in *The Permission to Be Right*.

About the Cover

The cover depicts Nīlakaṇṭha Somayājī's geo-heliocentric model. Five planets — Mercury, Venus, Mars, Jupiter, Saturn — orbit the Sun; the Sun, carrying them, orbits Earth. Nīlakaṇṭha arrived at this arrangement in Kerala in the early 1500s, roughly eighty years before Tycho Brahe proposed the same structure in Denmark. Mars is drawn with two paths: the solid circle is what the model predicted; the dashed ellipse, exaggerated for visibility, traces what Tycho's observations would actually find. The eight arcminutes between them are the residual Kepler refused to absorb.

www.ingramcontent.com/pod-product-compliance
Lightning Source LLC
LaVergne TN
LVHW100314210926
848480LV00003B/1113